ROUTLEDGE LIBRARY EDITIONS:
GEOLOGY

Volume 22

LANDSCAPE PROCESSES

LANDSCAPE PROCESSES

An Introduction to Geomorphology

DARRELL WEYMAN AND VALERIE WEYMAN

LONDON AND NEW YORK

First published in 1977 by George Allen & Unwin (Publishers) Ltd

This edition first published in 2020
by Routledge
2 Park Square, Milton Park, Abingdon, Oxon OX14 4RN

and by Routledge
52 Vanderbilt Avenue, New York, NY 10017

Routledge is an imprint of the Taylor & Francis Group, an informa business

British Library Cataloguing in Publication Data
A catalogue record for this book is available from the British Library

ISBN: 978-0-367-18559-6 (Set)
ISBN: 978-0-429-19681-2 (Set) (ebk)
ISBN: 978-0-367-31327-2 (Volume 22) (hbk)
ISBN: 978-0-367-31332-6 (Volume 22) (pbk)
ISBN: 978-0-429-31631-9 (Volume 22) (ebk)

LANDSCAPE PROCESSES

An introduction to geomorphology

Darrell and Valerie Weyman

London
GEORGE ALLEN & UNWIN
Boston Sydney

First published in 1977

ISBN 0 04 551026 1

Printed Offset Litho in Great Britain
in 10 on 11 point Times
by Cox & Wyman Ltd.,
London, Fakenham and Reading

Contents

	PREFACE	*page* 8
Chapter 1	LANDSCAPE PROCESSES	9
	A. Denudational processes	10
	B. Geological structure	10
	C. Time	12
Chapter 2	HUMID LANDSCAPES	13
	A. Water in the landscape	13
	B. Weathering	20
	C. Transportation on hillslopes	25
	D. Transportation, erosion and deposition in rivers	33
	E. Denudation in the drainage basin as a whole	38
	F. The special case of limestone erosion	44
Chapter 3	ARID AND SEMIARID LANDSCAPES	50
	A. Water in an arid region	50
	B. Weathering in arid regions	51
	C. Hillslope and channel processes in arid landscapes	52
	D. Landforms produced by water-based processes	52
	E. Transportation, erosion and deposition by wind	54
Chapter 4	GLACIAL AND PERIGLACIAL LANDSCAPES	58
	A. Water and ice in a valley glacier	58
	B. Processes of glacial erosion and transportation	60
	C. Landforms produced by glacial erosion	62
	D. Landforms produced by glacial deposition	64
	E. Glacial meltwater	67
	F. Ice-sheets	68
	G. Periglacial regions	68
Chapter 5	COASTAL LANDSCAPES	71
	A. Currents, tides and waves	71
	B. Marine erosion	72
	C. Coastlines where erosion is dominant	73
	D. Coastlines where deposition is dominant	76
	E. Deposition by rivers in coastal areas	80
Chapter 6	LANDSCAPES OF THE PAST	81
	A. Movements of land and sea	81
	B. Changes in climate	85
	FURTHER READING	91
	INDEX	93

Preface

The science of geomorphology (the study of the landscape) has in the last few years been through something of a revolution. Originally geomorphology was largely a description of landforms but, since it has become possible to actually measure the processes which create and erode the landscape, the subject has developed far more sound theory to explain how the landscape evolves.

At the moment, most school books in this area still tend to emphasise the range and diversity of landforms. In this book, we have deliberately attempted to develop theory at the expense of numerous examples. Furthermore, we have spent some time on the subject of humid environments because we believe that all landscape systems work in basically the same way and it is worth making an attempt to really understand the home environment first. In the same context, we have omitted the subject of landscapes created by volcanic activity and earth movements because these topics are dealt with at length in many other books. On the other hand, we make no apologies for including sections on hydrology and hillslopes, subjects often ignored in school, since we feel that some understanding of these items is essential in any wider appreciation of the forces at work in the landscape. The choice of photographs was made on the same basis: there will be no difficulty in finding pictures of glacial valleys but photographs of mass movement are more difficult to come by.

This is only a short book and it cannot therefore be more than an introduction to this wide and fascinating subject. In particular, we have not been able to include any detail on the methods of measurement used to carry out experiments in the landscape, nor have we felt able to develop the theme of practical application of the theory to problems of human concern such as soil erosion or slope stability. Suggestions for further reading in both of these areas are made at the end of the book. Suggestions for further reading of theory are also made at the end since some students of Advanced Level Geography may hopefully find the book a helpful next step from Ordinary Level but will undoubtedly wish to pursue the subject beyond the limits set in this volume.

Darrell and Valerie Weyman
Turnworth

Chapter 1

Landscape Processes

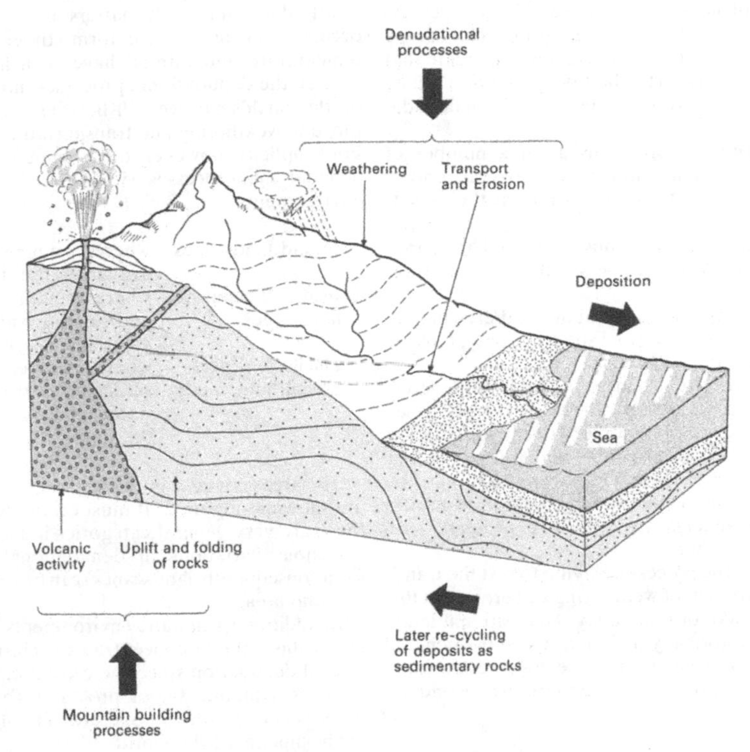

Figure 1 The geological cycle.

The landscape of the earth's surface is largely the creation of two sets of natural forces which act constantly against one another.

On the one hand, the apparently solid surface of the earth is little more than a thin crust overlying a molten interior. Movements within that interior can cause the injection of molten material into the solid outer crust, where it forms new rock, or the spilling of molten material on to the surface to create volcanoes and lava flows. This **volcanic** activity is supplemented by **earth movements**, which are the large-scale uplifting, downwarping and folding of the solid crust by interior earth forces. Earth movements are capable both of forming new rock (from uplifted marine sediments, for example) and of creating new landforms from existing rocks.

On the other hand, arrayed against these

internal forces of rock and landform creation are the external forces of weather, water, ice and sea which modify and destroy those rocks and landforms. This second group of forces is referred to as **denudation processes**. With the exception of a few features such as volcanoes, which owe their form largely to internal processes, most of the earth's landscapes are of denudation origin; that is, landscapes in the process of being destroyed.

Geomorphology is the science of landscapes. As a science it includes both a description of landscape and the analysis of the processes which create that landscape. In this particular book, attention will be focused very largely upon the processes of denudation.

Denudation actually consists of a number of separate operations, and it helps overall understanding to define these operations at the outset.

Weathering is the breakdown of solid rock into fragments which may be small or soluble fragments.

Transportation is the removal of weathered material away from its original site by gravity, moving water, ice or wind.

Erosion in the strict sense implies simultaneous weathering and transport by some processes such as rivers. For example, a river can cause the disintegration of rock in its bed and carry away the products at the same time. The term 'erosion' is also used more loosely to imply transportation in general since the two processes are often inseparable.

Deposition is the process of laying to rest the transported products of weathering and erosion in the sea, in a lake or in a valley. Depositional landforms are normally included under the general heading of denudation since they result from denudation proper and are therefore secondary landforms.

Denudation as a whole forms part of what is known as the **geological cycle**. If left undisturbed, deposits of weathered material will gradually form new sedimentary rocks as they are compressed by overlying material or cemented by circulating waters containing dissolved chemicals. If these rocks are later uplifted in earth movements, they will form new landscapes on which denudation can again operate (Fig. 1).

All denudational landscapes can be viewed as the result of the interaction between denudational processes and the underlying geological structure and the time during which that interaction has been taking place.

A. Denudational processes

The processes of weathering, transportation, erosion and deposition are not constant throughout the world but vary with changes in the surface environment. In particular, climate will quite clearly determine such matters as whether water occurs in normal or ice form. Indeed, because temperature and rainfall have such fine control over all the denudational processes, no two places on the earth's surface will have exactly the same range of weathering and transportation processes. For simplicity, however, this volume will consider denudational processes within three basic climatic environments:

1. **humid** landscapes—where running water is available at all times;
2. **arid** and **semiarid** landscapes—where rainfall is very low and running water is available only intermittently;
3. **glacial** and **periglacial** landscapes—where temperatures are sufficiently low to maintain water as ice for much of the time.

The areas covered by these three environments are shown in Figure 2. It must be emphasised that these are very general categories indeed and that enormous variations in denudational processes (and consequently landscapes) can be found within any one area.

In addition to climatic environments, one chapter in this volume concentrates exclusively upon coastal denudation since the coastal environment involves a unique set of processes (based upon wave action) which operate in all climatic environments of the world.

B. Geological structure

The role of denudational processes is to create new landscapes from the initial forms created by volcanic activity and earth movements. Throughout the life of the denudational landscape, evidence will remain of its original form as a geological structure.

Geological structure influences denudational landscapes through both the nature of the rocks

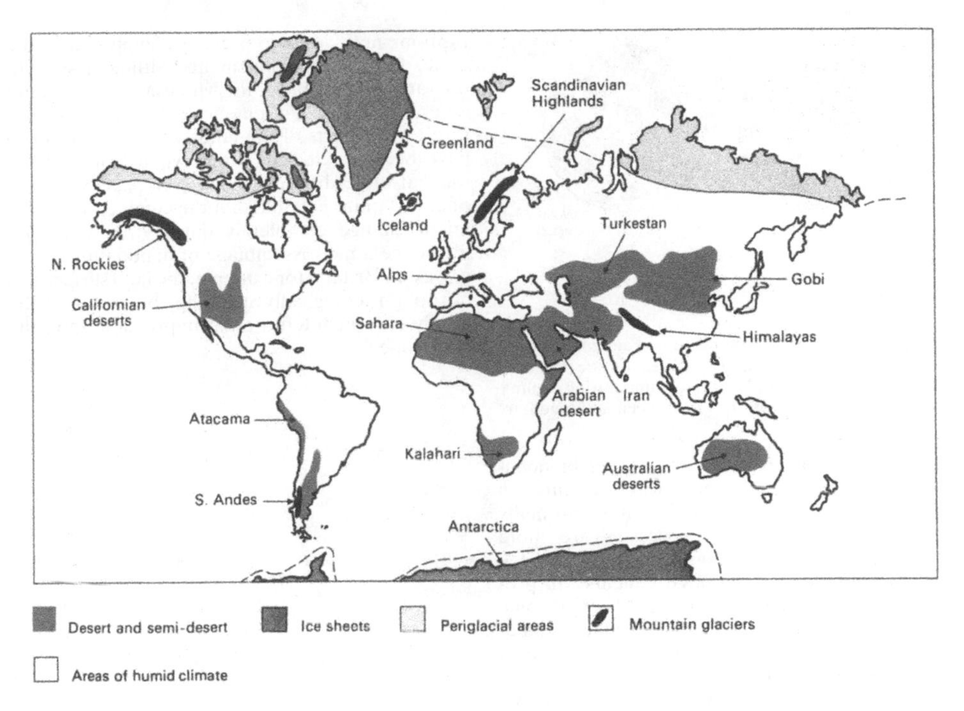

Figure 2 Areas of distinctive denudational processes.

concerned and the forms in which those rocks occur.

(i) Rock type. Some types of rock are better able to resist weathering and erosion than others. Generally speaking, **igneous** rocks such as granite and basalt, which are produced by volcanic activity, consist of minerals which have grown together in very tight bonds during the cooling of the original molten material. Such rocks are frequently more **resistant** than the **sedimentary** rocks formed in the geological cycle, which consist of weathered particles compressed or cemented together (Fig. 3). Obviously there is a vast range of resistance even within the sedimentary rocks—thus, heavily cemented quartz-sandstones may prove highly resistant in comparison with weakly bonded clay rocks.

All other things being equal, the more resistant rocks form upstanding features within a denudational landscape.

(ii) Rock form or structure. The form in which rock was left by volcanic activity or earth movement may continue to influence the shape of the denudational landscape. Volcanic activity may produce volcanoes at the surface or masses of intruded igneous rocks beneath the surface which are later exposed by denudation—thus, the overall form of Dartmoor is determined partly by the shape adopted by the granite when it was intruded into the earth's crust. Similarly, earth movements may uplift and fold rocks into many patterns. The Pennines consist of a long upfold (anticline) of rocks with the axis running north–south and the highest ground found on the ridges of more resistant gritstone towards the centre of the fold. Much of south-east England is dominated by a sequence of scarplands (limestone and chalk) with intervening clay vales, the whole being part of a sedimentary sequence which has been tilted down towards the south-east.

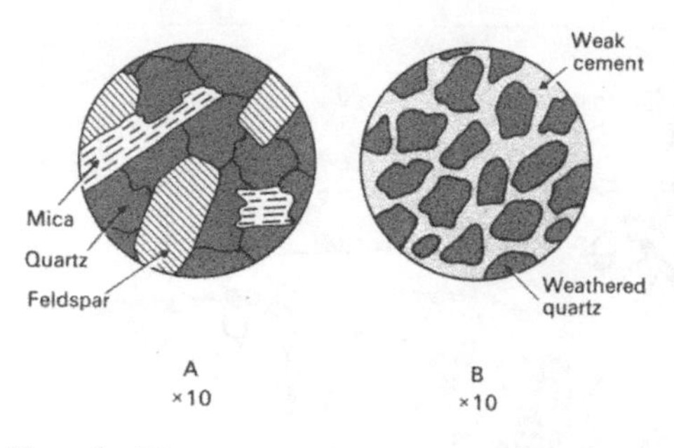

Figure 3 Microscope sections of (A) an igneous rock, granite and (B) a sedimentary rock, sandstone. Minerals in the igneous rock have grown together during cooling, those in the sedimentary rock have been laid together.

In assessing the relative importance of denudational process and geological structure in producing *distinctive* landforms, it is normally argued that denudational processes are more significant. It is usually assumed that the same rock in different denudational environments will produce markedly different landforms. There is some argument that a few rock types produce a landscape which is recognisable in any denudational environment. To some extent this may be true for granite, and it is probably more true for limestone. For this reason a short section on limestone landscapes is included at the end of Chapter 2.

C. Time

The longer the period during which denudational processes operate upon a given geological structure, the greater will be their effect. Generally, these processes operate slowly or at very irregular intervals and, although it may be possible to observe processes such as the transportation of sediment by a river, only minor changes in the shape of the river channel are likely to be seen over several years. Considerably longer periods are required before major changes can be seen in a river valley, but given time, the denudational processes will gradually wear away the landscape to something resembling a relatively flat plain. Obviously, the time taken to achieve total denudation will vary with the geology of the underlying rocks, but the time scales involved are probably of the order of several million years.

On the whole, it seems likely that this denudational 'cycle' is rarely completed. Two major interruptions may occur to prevent completion of the cycle. First, earth movements, although slow to operate, are capable of the wholesale uplift of land areas over this sort of time scale. It is therefore not rare to find parts of the landscape which seem to have been created at a time when land and sea were at quite different levels (see page 81). Second, it is quite possible that the climate of an area might change completely during that time and introduce a new assemblage of denudational processes. In Britain, for example, the land surface has not long (geologically speaking) been free of the ice sheets which left a great impression upon the landscape.

Chapter 2

Humid Landscapes

A. Water in the landscape

Water is of vital importance to the denudational processes of the humid landscape. Not only does it take part in the weathering of rock, but it also forms one of the main methods of transporting weathered material. It is therefore necessary to examine how water moves in the landscape in order to understand how denudation takes place.

1 *The water cycle*

The cyclic movement of weathered material has already been considered briefly (see Fig. 1). Water also moves in a cycle, although the time taken to complete the cycle is likely to be measured in weeks rather than millions of years.

Water enters the landscape as **precipitation** (rainfall or snowfall). It may then enter soil, plants or rock, but it eventually leaves the landscape by one of two routes. It may either form **runoff** (riverflow or iceflow) and so make its way back to the ocean, or it may be passed directly back to the atmosphere as water vapour due to **evaporation** from standing water or **transpiration** from the leaves of plants which have drawn water from the soil. The water which is lost from the river basin as runoff passes through the oceans before being eventually evaporated back into the atmosphere. Water vapour from either land or ocean will condense to form the new precipitation which completes the cycle.

2 *Water routes in the landscape*

Over much of the landscape, the ground is covered by a thick mat of vegetation which **intercepts** (catches) precipitation and holds it for a short time. During light rainfall, all the water may be intercepted and then evaporated back into the atmosphere, but usually water starts dripping through the vegetation and on to the surface of the soil.

Soil consists of solid particles of mineral matter (from the underlying rock) together with pieces of organic matter such as plant roots and dead leaves. Between the solid particles are spaces which can be occupied by water. Water moves into these spaces by the process known as **infiltration**. The rate of water infiltration into the soil depends upon two things: the sizes and shapes of the air spaces in the soil, and the amount of water already filling those spaces.

1. The total volume of air spaces in a soil is the soil **porosity**. If, for example, a 100 cm³ block of soil consisted of 60 cm³ of solid particles and 40 cm³ of air space, the porosity would be 40%. Generally speaking, the porosity of a soil increases as the average size of particle decreases (Fig. 4). This occurs because the packing of very small particles leaves more air spaces over all.

Infiltration depends less on the total amount of air space (porosity) than upon the *size* of the air

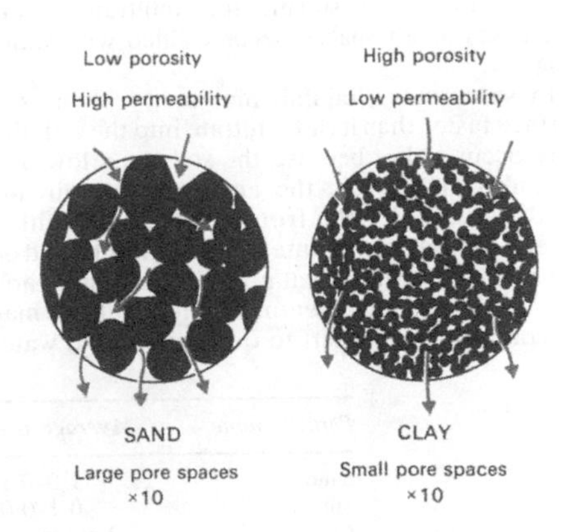

Figure 4 Porosity and permeability in soil.

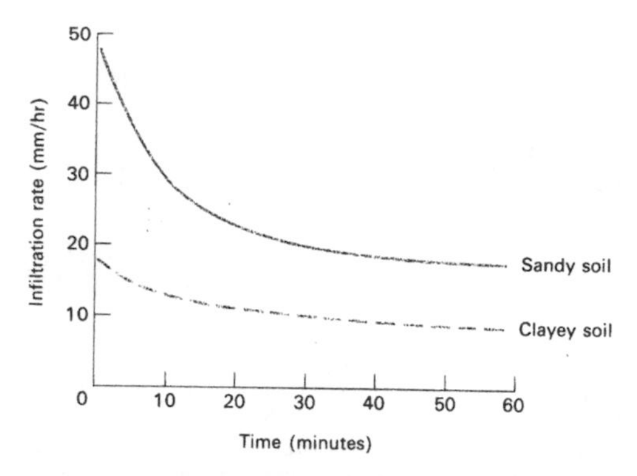

Figure 5 Contrasts in the rate of water infiltration in two soils.

spaces, because where they are very small, water becomes attracted easily to the surrounding particles and moves more slowly. The **permeability** of a soil (the rate at which water moves through the air spaces) may therefore increase as porosity decreases. Consequently, the rate of water infiltration is usually highest in soils with either sand-size particles (Fig. 5) or in soils with additional air spaces made by plant roots, insects and worms.

2. The rate of infiltration will also vary with the proportion of the available air spaces which are already occupied with water. It will be obvious that the infiltration rate of a dry soil is higher than that of the same soil after rainfall. During the course of a storm, the infiltration rate decreases as air spaces become filled with water (Fig. 5).

In some cases, rainfall may arrive at the soil surface faster than it can infiltrate into the soil. this may occur either because the soil has a low permeability or because the air spaces are already partly filled with water from earlier rain. In either case, if rainfall cannot enter the soil, it will stand on the surface forming small pools in the minute hollows of the ground. Eventually, these hollows may fill completely and start to overflow so that water begins to run downhill over the ground surface as **overland flow** (also referred to as **surface runoff**). Once overland flow has started, rainfall will be carried downslope to the river very quickly, causing a rapid increase in riverflow. This rapid increase in riverflow is supplemented by rain falling directly into the river itself as **channel precipitation**.

Generally speaking, overland flow on a large scale is rare in the British Isles except during very intense storms. The reasons seem to be that rainfall intensities are usually low in this country while most soils have very high infiltration capacities. The exceptions to this rule are those areas of slopes which tend to remain wet because of subsurface drainage (see below). At the base of hillslopes and in hollow areas where water is concentrated, overland flow may occur even under light rainfall conditions.

Water which *has* infiltrated will move slowly downwards through the soil under the pull of gravity. Some of this water will be drawn out of the soil by plant roots, eventually to be transpired through the leaves of the plant and evaporated into the atmosphere, but most of the water will continue to move downwards. The lower layers of a soil are usually compacted by the weight of the overlying soil, with the result that permeability decreases towards the bottom of the soil. Consequently, the vertical movement of water becomes slower and, if the soil is on a slope, water will be deflected downslope as **throughflow** (Fig. 6). Throughflow is usually much slower than overland flow and it may be days or weeks before rainfall reaches the river by this process. Nevertheless, throughflow is important in providing riverflow in periods without rainfall and in maintaining wet soil conditions at the base of the slope, which encourage overland flow during subsequent storms. Occasionally, the action of throughflow causes the development of underground **pipes** in the soil. These pipes, which may be anything from a few centimetres to almost a metre in diameter, can carry water underground at much higher speeds than throughflow. In Britain, pipes seem

Particle name	*Average size* (mm)	*Typical porosity* (%)
Sand	1·0–0·1	30–40
Silt	0·1–0·002	25–45
Clay	Less than 0·002	50–60

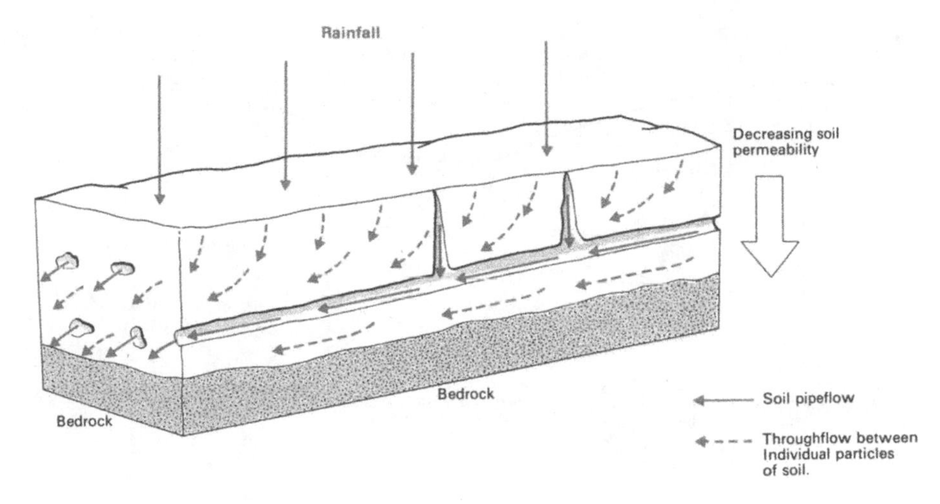

Figure 6 The movement of water downslope within the soil.

to be most common in upland areas with heavy rainfall and steep slopes.

The rock beneath a soil is usually less permeable than the soil, with the result that most water moves downslope within the soil as throughflow. In some cases, however, the rock itself is permeable because spaces exist between the rock particles. Both sandstone and chalk are usually permeable. Limestone and basalt are also permeable, but this is due to the large number of joints and cracks in the rock, the solid rock being more or less impermeable. Having passed through the soil, water will enter permeable rock and **percolate** vertically downwards. At some considerable depth beneath the surface, the weight of overlying material causes the permeability of the rock to decrease (Fig. 7). Above this point, water, being unable to move further downwards, collects and fills the rock until it is **saturated**. The part of the rock which is saturated is referred to as the **ground-water zone**. The upper surface of the ground-water zone is the **water table**. Above the water table, the rock is unsaturated and water percolates vertically under the pull of gravity. Within the ground-water zone, however, pressures caused by the weight of overlying water result in water moving towards the river by complex routes. At some points in the ground-water zone, water may actually be forced to flow *upwards* to the river channel from some depth (Fig. 7). Ground-water zones store very large quantities of water and can supply rivers with water for months after rainfall.

Ground-water usually emerges along the bed of a river and cannot easily be detected except by noting that riverflow increases downstream even when tributary streams do not enter. In some situations (Fig. 8), a permeable rock is underlain by an impermeable one such as clay or shale. Under these conditions, ground-water in the permeable rock may be forced to emerge as **springs** above the impermeable layer. It must be remembered, however, that springs of this type are not the major source of riverflow. Riverflow is produced mainly by throughflow from the soil, or ground-water flow from the rock, moving directly into the channel below the water surface of the river.

3 *The drainage basin*

The amount of water flowing down a river (riverflow or **discharge**) depends mainly upon the size of the area of land supplying water and second upon the precipitation. The area supplying water to a point on a river is called the **drainage basin** or **catchment** for that point. The drainage basin for a point where discharge is measured can be found from a map (Fig. 9).

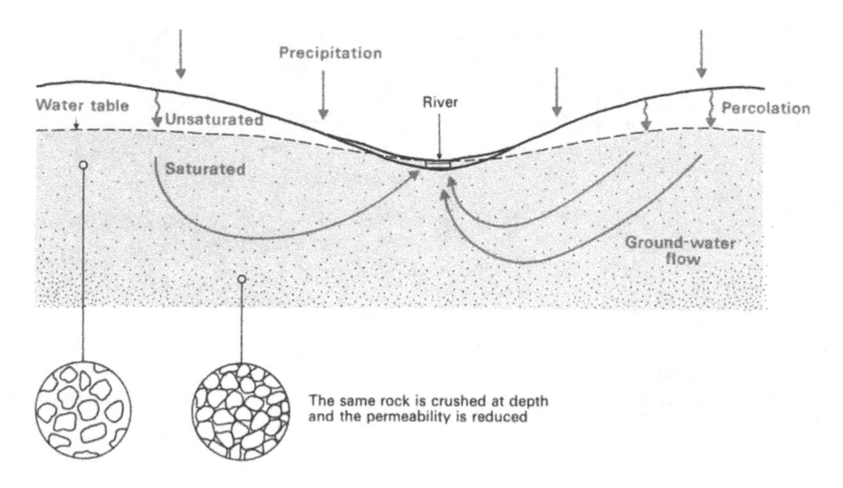

Figure 7 The movement of water within solid rock.

The boundary, or **watershed**, of the drainage basin is a line following the maximum slope up from the measurement point. On a map, this line crosses all contours at right angles. If the boundary is drawn correctly, all rainfall landing within the basin is capable of flowing into the river above the measurement point.

For the drainage basin as a whole, precipitation is expressed in millimetres as the average depth to which water would stand if it did not soak into the ground or run downslope. Evaporation and transpiration (usually referred to jointly as **evapotranspiration**) and runoff are also expressed in millimetres as the relevant proportion of total precipitation. Although the total *volume* of water produced by a river will increase with the size of the drainage basin, by expressing runoff as a proportion of precipitation depth it is possible to compare the effects of precipitation and evapotranspiration upon the runoff of basins of different sizes.

4 *The water balance*

Water enters a drainage basin as precipitation and leaves it again as evapotranspiration or river discharge (runoff). During the course of one year a drainage basin usually becomes neither increasingly wet nor increasingly dry. The basin will remain in roughly the same moisture state. In other words, the amount of water entering the basin must be very close to the amount leaving. That is:

input of water to the basin = output of water from the basin

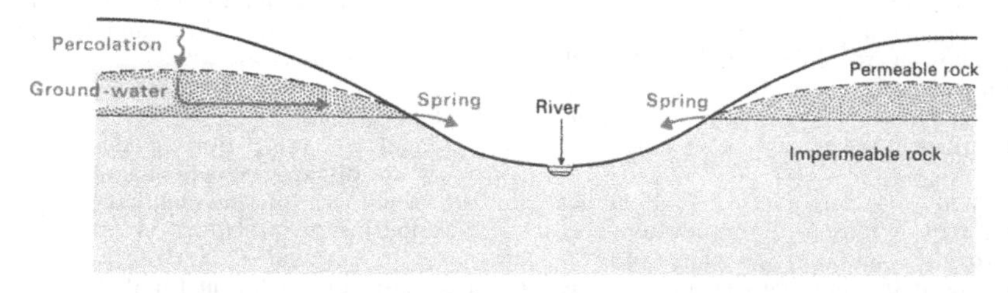

Figure 8 The formation of springs where a permeable rock is underlain by an impermeable rock.

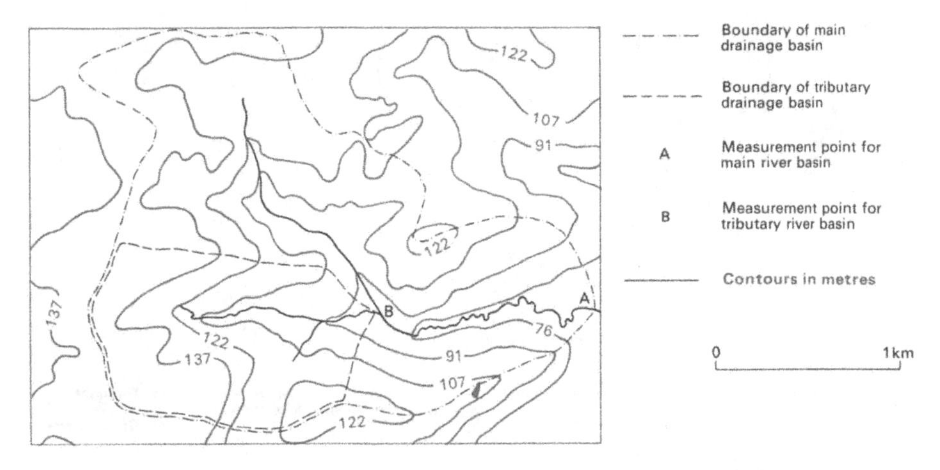

Figure 9 A small drainage basin within a larger drainage basin: headwaters of the River Chew, North Somerset.

This simple idea can be expressed as the **water balance equation**:

precipitation = runoff+evapotranspiration

Even in Britain the range of values in the water balance equation is considerable. Extreme values are provided by the Rivers Ystwyth and Yare:

	Precipi-tation		*Runoff*		*Evapo-transpira-tion*
R. Ystwyth	1563 mm	=	1289 mm	+	274 mm
R. Yare	739 mm	=	132 mm	+	607 mm

The River Ystwyth is located in Wales, where rainfall is heavy, owing to altitude, and evapotranspiration low, owing to persistent cloud cover and low temperature, with the result that the annual runoff is high and represents almost 80 % of the total precipitation. In contrast, the River Yare, in East Anglia, has a much lower precipitation and a correspondingly low runoff. Moreover, higher temperatures and less cloud lead to high evapotranspiration rates, so that runoff is less than 20 % of the total precipitation.

So far, it has been assumed that there is a balance between the input and output of water for the drainage basin because there is no change in the moisture state of the basin. Strictly speaking, of course, there are bound to be slight variations. Even if the basin is not becoming progressively wetter or drier, it may happen to be slightly wetter at the end of a year than it was the beginning. A storm occurring on the last day of a year, for example, will provide water which is still being **stored** in soil and rock *en route* to the river (Fig. 10) at the time when the measurement of river discharge stops. Consequently, the water balance equation

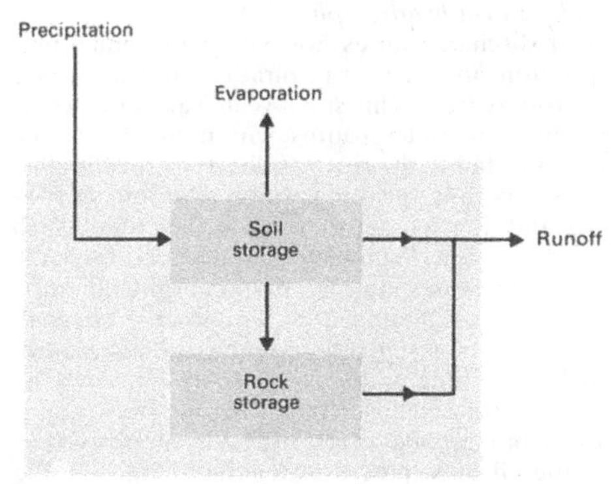

Figure 10 How water is stored between entering and leaving a drainage basin.

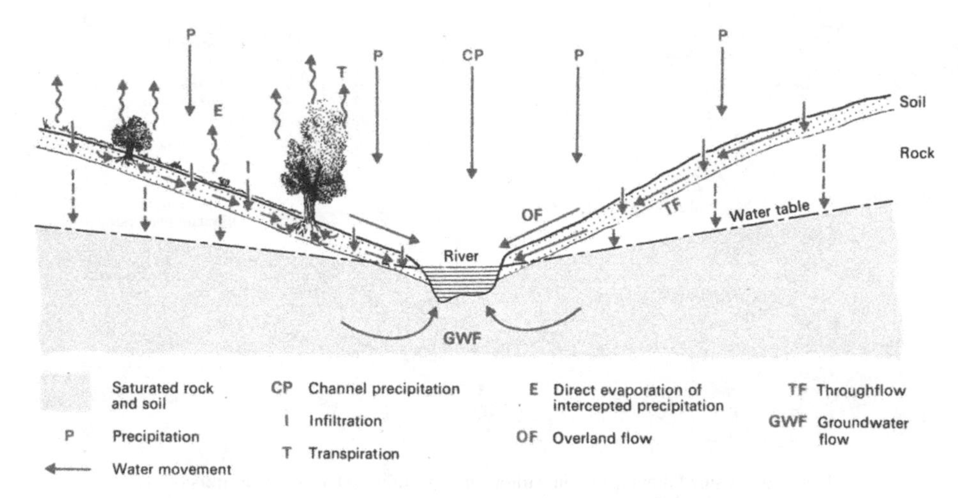

Figure 11 The various routes taken by water within a drainage basin. The speed of water movement decreases with depth below the surface.

should be altered to allow for slight variations in the amount of water being stored in the basin between the beginning and end of the year:

$$\text{precipitation} = \begin{array}{l}\text{runoff} + \text{evapotranspiration} \\ \pm \text{change in storage}\end{array}$$

Over one whole year, the change in storage will be small in comparison with the total amount of precipitation.

5 *The storm hydrograph*

River discharge varies not only with annual precipitation and evapotranspiration, but also on a day-to-day basis. This small-scale variation occurs because the water routes within the landscape carry rainfall to the river at different speeds (Fig. 11). Generally speaking, surface runoff such as overland flow is faster than subsurface runoff such as throughflow. Following rainfall, river discharge usually increases rapidly as channel precipitation and overland flow feed water into the channel. These processes stop soon after rainfall has ceased, but by this time, water is being brought into the river by throughflow. When throughflow stops, flow from ground-water will still supply water. Taking all these processes together (Fig. 12), the river discharge following rainfall usually shows a rapid rise and then a fall which is dramatic at first but later becomes a long-drawn-out curve when plotted as a graph. This distinctive pattern is called the **storm hydrograph** of the river.

During the course of a year, river discharge shows a pattern of separate storm hydrographs with intervening periods of low flow, that is **baseflow**. As can be seen from a typical section of riverflow record (Fig. 13), the height of each storm hydrograph is related to the amount of rainfall plus the amount of water from previous rainfalls still being stored in the basin, the latter factor controlling how much rainfall can infiltrate into the soil.

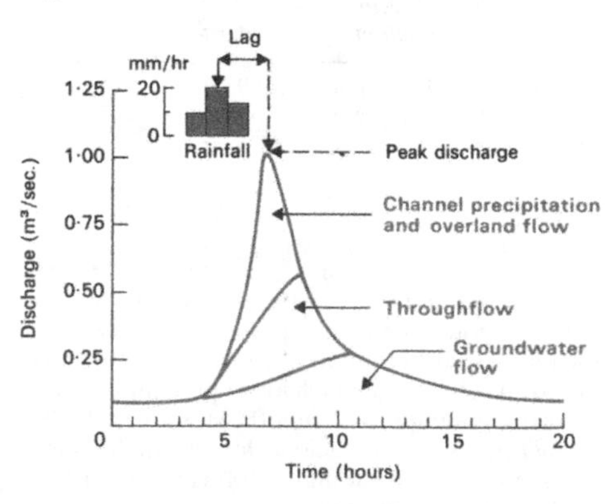

Figure 12 The storm hydrograph for a river, showing the times at which various runoff processes feed water into the river channel.

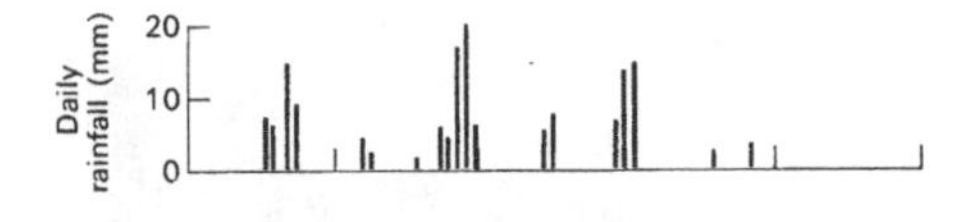

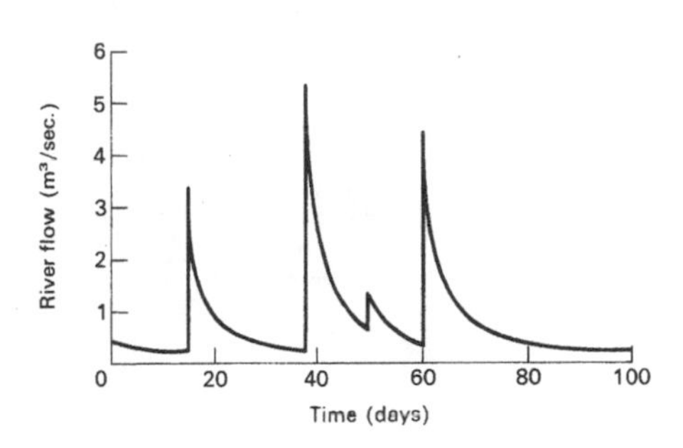

Figure 13 The pattern of rainfall and riverflow for a river basin.

Very large storms are rare but smaller peaks occur quite frequently.

Information of this sort is used to construct a **flow duration curve** (Fig. 14). Using Figure 13 as an example of 100 days' record from a river (normally a record of many years is used), it is possible to calculate the percentage of the total time during which a given value of discharge is exceeded. For example, on only 1 day out of the 100 did the discharge exceed 5 m^3/s. This is plotted on the curve as 5 m^3/s being exceeded 1 % of the time. Similarly, a discharge of 4 m^3/s was exceeded on 2 days (2 %), 3 m^3/s on 4 days (4 %), etc. It is possible to work on down the discharge scale as far as 0·2 m^3/s, where the discharge was exceeded on every day (100 %). The complete flow duration curve is extremely valuable because from it an estimate of how often a discharge might occur can be made. For example, a discharge of at least 3 m^3/s on 4 days out of every 100 might be expected. With a riverflow record of many years, it might be possible to assess quite accurately the low flows on which to base water supply planning or the high flows on which to plan flood control.

6 *Differences between drainage basins*

In section (4) above the Rivers Ystwyth and Yare were compared on the basis of annual precipitation and evapotranspiration. It is also possible to find drainage basins where the total amount of runoff is the same but the pattern of runoff during the year is different. The Rivers Avon and Kennet in Wiltshire, for example, both gave about 160 mm of runoff in 1964. As Figure 15A shows, however, although the pattern of monthly rainfall was very similar, the runoff from the River Avon was much more variable than that from the River Kennet. In other words, the River Avon responds more quickly to rainfall and falls more drastically during dry periods than does the River Kennet. These contrasts occur because the River Avon is supplied largely by rapid runoff processes, whereas the River Kennet is more reliant upon slower subsurface processes (mainly ground-water flow). In this case, the processes differ because the River Avon is largely underlain by clay rocks which are impermeable and encourage the development of overland flow, while the chalk rock of the Kennet basin takes most of the rainfall into the ground-water zone and releases it slowly (Fig. 15B).

Precisely the same effects can be found where there are differences in soil type or land use. In Figure 16A, for example, a small basin in the Mendip Hills of Somerset is shown to include two types of soil. In the upper part of the basin, the top of the soil is a peat layer with a low infiltration capacity, whereas farther down the basin a very sandy soil with a high infiltration capacity occurs on the steeper slopes. The contrasting hydrographs for the two parts of the same basin (Fig. 16B) show that overland flow on the peat soil gives a very sharp runoff curve, while in the lower basin water is reaching the stream by throughflow from the side slopes and the hydrograph is consequently much spread out over time. As far as land use is concerned, forested areas usually show fairly

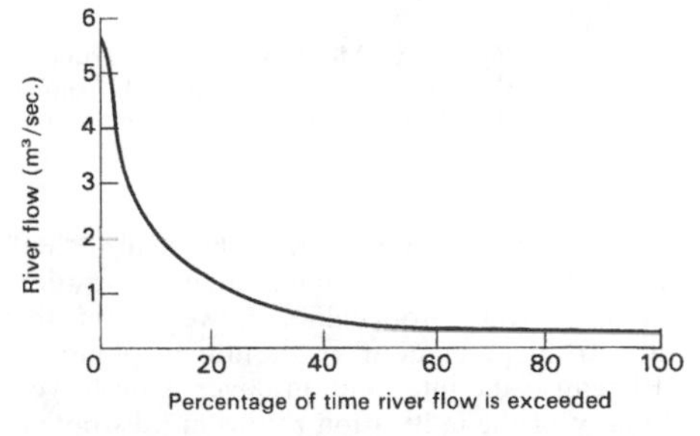

Figure 14 The flow duration curve for the riverflow record shown in Figure 13.

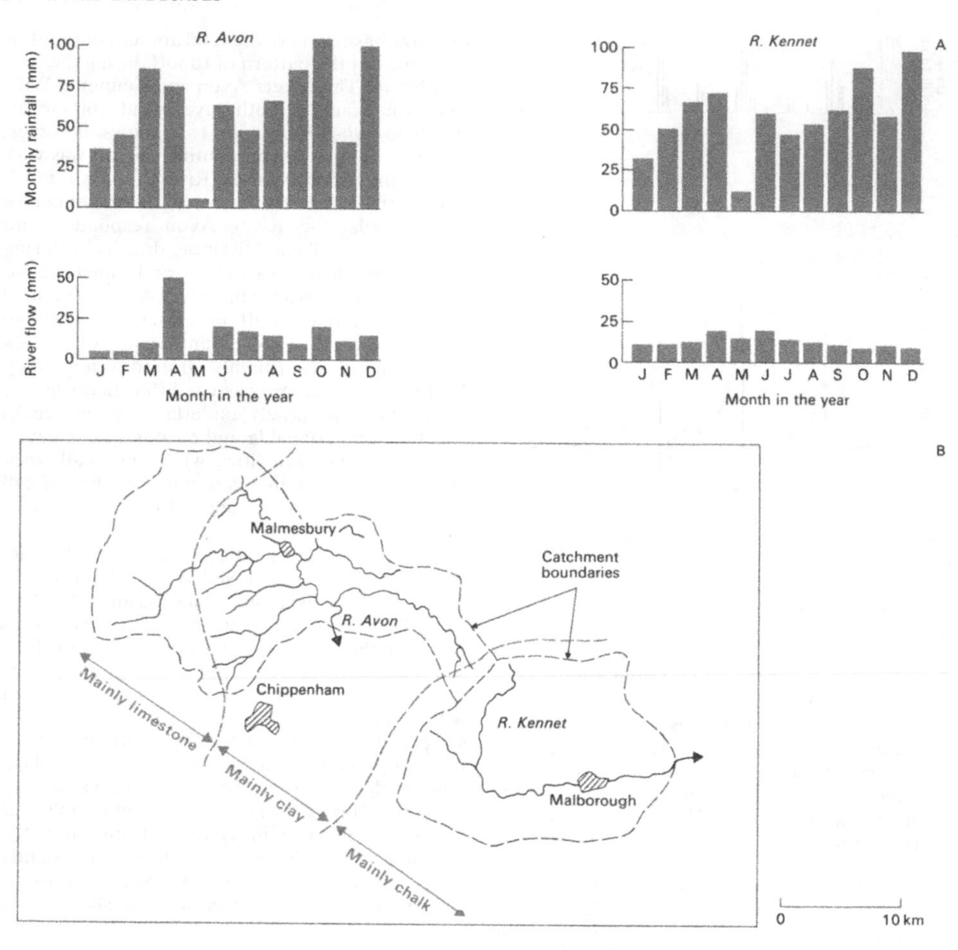

Figure 15 Monthly rainfall and runoff graphs for the Rivers Avon and Kennet (A), showing that, although rainfall is similar, runoff patterns are different because the River Avon is partly underlain by impermeable clay rocks (B).

steady runoff in comparison with pasture land because tree roots maintain high infiltration capacities. More noticeable, however, is the change from permanent vegetation to ploughed land because the interception cover is destroyed and many of the infiltration routes are disrupted. Overland flow tends to be common on freshly ploughed land.

B. Weathering

Weathering is one of the main elements in the denudational cycle. Solid rock cannot normally be transported from its place of origin unless it is first broken down by weathering into manageable fragments or soluble chemicals.

It is helpful to distinguish between **physical weathering**, which simply breaks the rock down

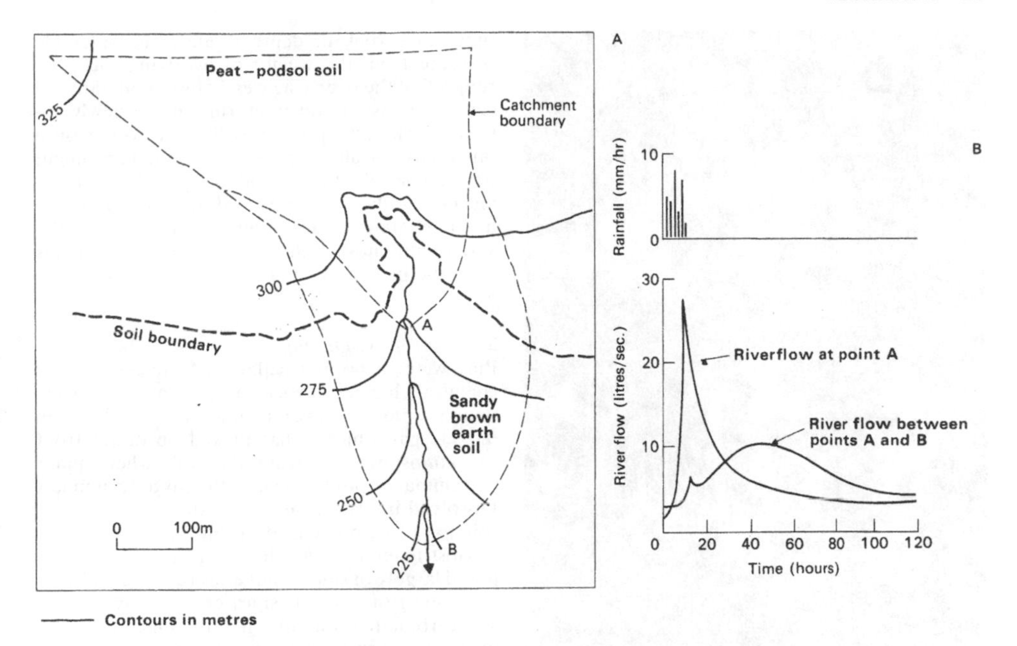

Figure 16 A small drainage basin in the Mendip Hills, Somerset (A) showing that a single storm produces a different river hydrograph (B) in two parts of the same basin because the soil is more permeable in the lower part of the valley.

into smaller fragments, and **chemical weathering**, which causes a chemical alteration and breakdown of the rock material. Generally speaking, physical weathering involves fairly large pieces of rock at one time, whereas chemical weathering takes place on a microscopic scale and involves individual atoms or molecules. Although these two types of weathering are here considered separately, it is important to notice that they normally take place together in the landscape.

1 *Physical weathering*

Any piece of rock will eventually disintegrate if a sufficiently strong force is applied. In the natural environment, forces of this sort can be created by water freezing, by changes in water content, and by the activity of plant roots. In nearly all these cases, the processes are aided by the existence of cracks or joints in the rock. Consequently, a rock such as shale, which has many fracture lines, is always more easily weathered than a massive rock such as sandstone, which has very few natural cracks.

(*a*) *The freezing of water.* Most substances contract when they are cooled. Water is a little unusual in this respect since between freezing point (0 °C) and −4 °C ice actually expands. If water is lodged in a crack in a rock and then freezes, it will expand and press against the walls of the crack, causing the rock to break. This process is most effective not in glacial regions, where water is permanently frozen, but in areas of the world where the temperature range causes alternate freezing and thawing of water. Water can then work its way even farther into the rock, opening the crack a little more each time it freezes.

It is frequently suggested nowadays that the pressures involved may be created not so much by the expansion of ice below 0 °C but by the growth of the ice crystals themselves at the point of freezing.

(*b*) *Changes in water content.* Some rocks, and particularly those with a high clay content, expand slightly when wet and then contract again when

Photograph 1 Physical weathering by the growth of ice crystals on an exposed river bank.

dry. Repeated wetting and drying may cause fracturing to occur in the material. As with the freezing process, wetting and drying is most effective in areas where the moisture conditions are constantly changing. The process is most important in climates with distinct wet and dry seasons.

(*c*) *Plant activity*. Plant roots usually grow in soil, which is already-weathered material. Nevertheless, trees will sometimes extend their root systems into fresh rock beneath the soil where cracks already exist. The growing root can then exert a surprisingly powerful force on the further development of the crack, causing eventual rock disintegration.

(*d*) *Unloading of containing pressure*. It is thought that, since rock at depth is under considerable pressure from the weight of overlying rock, the removal of the overlying weight by denudation will cause expansion and fracturing in the newly exposed rock. This process is likely to occur on a fairly large scale: for example, fractures might develop parallel to the eroded sides of a river valley. Generally, the process is difficult to observe in humid climates as a weathered mantle cloaks the surface and most evidence for its operation comes from the bare rock areas of arid zones (see page 52).

2 *Chemical weathering*

Pure water is not particularly effective at causing chemical changes in rock-forming minerals. Most natural water, however, contains dissolved carbon dioxide gas which it has picked up either from the atmosphere or from the soil, where plant and animal respiration causes the gas to accumulate. Dissolved in water, carbon dioxide forms a weak solution of carbonic acid (Table 1).

Most igneous and sedimentary rocks are composed largely of silicate minerals (quartz, feldspar, mica, etc.) or hydrous silicates (the clay group). The carbonate group of minerals, which form the limestones, are the only major non-silicate rock-forming minerals. In general terms, a silicate in the presence of carbonic acid gradually breaks down. The products of that breakdown often include both a soluble component, which is then carried away by the water, and an insoluble residue which is left at

Table 1 *Examples of weathering equations*

(a) Formation of carbonic acid:

water + carbon dioxide → carbonic acid

$$H_2O + CO_2 \rightarrow H_2CO_3$$

(b) Breakdown of orthoclase feldspar:

orthoclase feldspar + carbonic acid + water →
potassium carbonate + kaolin clay + quartz

$$2K.Al.Si_3O_8 + H_2CO_3 + H_2O \rightarrow \underset{\textbf{soluble}}{K_2CO_3} + \underset{\textbf{insoluble}}{Al_2(OH)_2Si_4O_{10}.H_2O} + \underset{\textbf{insoluble}}{2SiO_2}$$

(c) Breakdown of calcium carbonate (limestone):

calcium carbonate + carbonic acid →
calcium bicarbonate

$$CaCO_3 + H_2CO_3 \rightarrow \underset{\textbf{soluble}}{Ca(HCO_3)_2}$$

the point of weathering until transported by some surface agency.

Not all minerals decompose at the same rate. Even among the common silicate minerals, it is possible to list the degree of resistance to chemical attack:

Least resistant	Olivine
	Augite
	Hornblende
	Mica
	Feldspar
Most resistant	Quartz

Since most rocks are composed of a range of minerals, including some from the list above, the rock itself may actually fall apart because of weathering of the less resistant minerals long before the more resistant have been attacked. Granite is composed largely of quartz, feldspar and mica. As chemical weathering takes place, the mica and feldspar largely turn to clay, leaving the quartz, in the form of sand, embedded in the clay (Fig. 17).

The clays (hydrous silicates) tend to be the end products of chemical weathering and are therefore not themselves very susceptible to further chemical change.

The best-known, but *not* the most common, chemical weathering reaction is the one between carbonic acid and a carbonate mineral found in limestone or chalk. The chemical reaction in this case is very straightforward (Table 1) and there is only one product of this reaction: calcium bicarbonate, which is soluble. It is this lack of insoluble residue (Photograph 2), other than that provided by impurities in the rock, which has led to the idea that limestone is particularly prone to chemical attack. Limestone is very prone to chemical attack, but so too are many other rock types which leave an insoluble residue.

The rate of chemical reaction varies with three main factors: the type of minerals present in a rock, the availability of natural acids, and temperature. The availability of natural acids increases partly with the flow of water but depends mainly upon the presence of carbon dioxide gas. From this point of view, chemical weathering *may* be more effective beneath some soil covers, where carbon dioxide content is increased by biological activity, than on bare rock surfaces. It should be noted incidentally

Photograph 2 A bare limestone slope subject largely to chemical solution

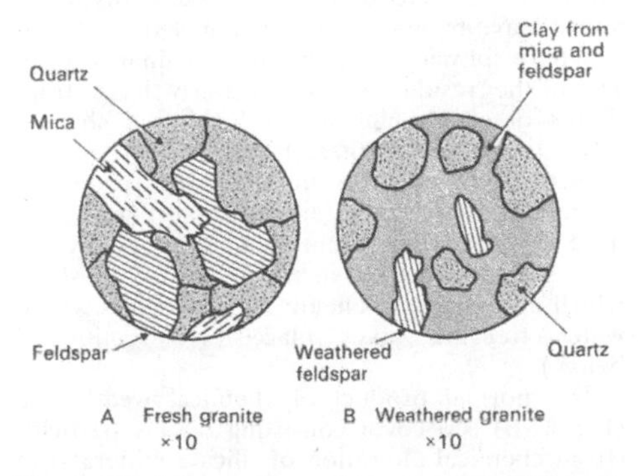

Figure 17 Microscope sections showing (A) fresh granite and (B) granite after some of the minerals have been altered by chemical weathering.

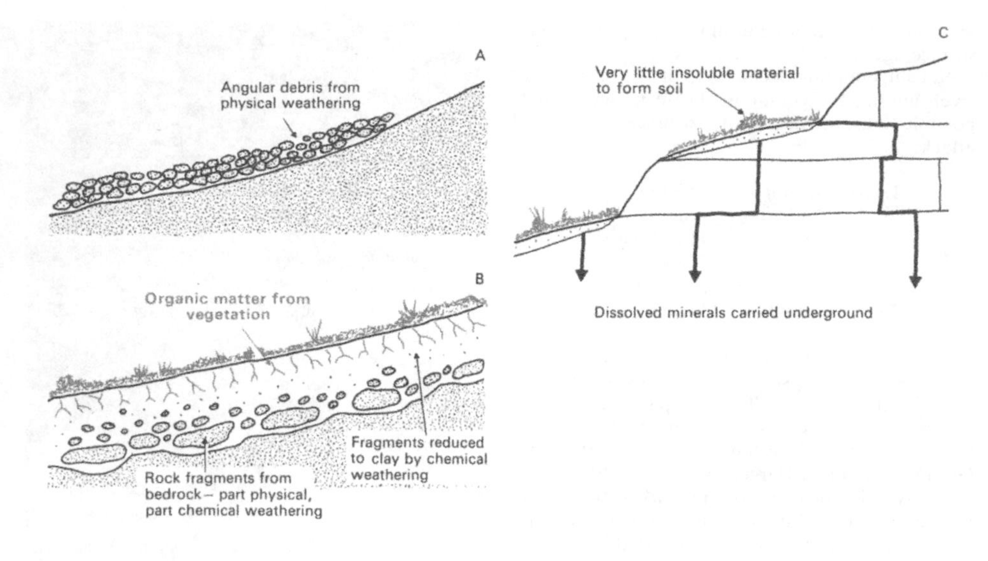

Figure 18 The products of weathering: (A) large rock debris from physical weathering; (B) a soil of mixed-size particles from physical and chemical weathering; (C) almost bare limestone where chemical weathering leaves little residue.

that organic acids derived directly from the decomposition of plants may also play a role in chemical weathering.

Most chemical weathering consequently occurs on hillslopes, although part of the total erosion effect of a river or the sea may include chemical action (corrosion).

As far as temperature is concerned, most chemical reaction rates increase with temperature with the result that weathering generally proceeds far more quickly in tropical humid areas than in temperate humid zones. There are exceptions to this fairly simple idea: limestone, for example, is found to increase in solubility at lower temperatures.

3 *The relative importance of physical or chemical weathering*

It is generally agreed that, taking the world as a whole, chemical weathering is more effective than physical weathering. Chemical weathering takes place almost everywhere, the main exception being the glacial regions, where water is frozen. Even in deserts, there is usually enough moisture to produce chemical reactions. Physical weathering is most effective either in areas of climatic extremes or in the limited areas of temperate regions where erosion agents such as rivers or waves operate. Over most of the temperate landscape, chemical weathering is likely to be the dominant method of rock breakdown.

4 *The products of weathering*

Most of the weathering processes discussed above leave behind a residue (Fig. 18).

Where physical weathering dominates (Fig. 18A), the residue is usually fairly large fragments of rock which are little altered chemically. Over bedrock, these fragments may form a debris cover which is unstable (i.e. prone to movement) since there are no small particles to hold the large fragments together. The commonest type of rock debris is the scree slope which accumulates beneath highland cliffs where water freezing takes place (see section C below).

The normal product of chemical weathering (Fig. 18B) is a cover consisting of clay particles (from chemical alteration of silicate minerals) in which are embedded unaltered fragments of more resistant minerals. If this material is colonised by vegetation, a soil develops which is a mixture of organic and mineral matter (Photograph 3). In most soils (Fig. 18B), the surface layers are dominated by organic matter and the bottom layers consist of larger fragments of bedrock loosened by chemical weathering along cracks. The middle layers of a soil consist of clay and mineral fragments in the process of weathering. The existence of a vegetation cover on the surface usually means that soil, unlike rock debris, is a fairly stable element in the landscape.

Because the chemical weathering of chalk and limestone leaves very little insoluble residue, only very thin soils form on these rocks (Fig. 18C). Limestone areas, in particular, are characterised by many bare outcrops of rock with no cover.

C. Transportation on hillslopes

Once solid rock has been broken down by weathering, it can be transported away from its original site. Although weathering takes place at different rates within the landscape, the final shapes of that landscape are due as much to variations in transporting processes as to variations in weathering rates. Of the transporting agents, the most obvious in a humid temperate climate are the rivers. Examination of the landscape (Fig. 19), however, indicates that rivers may actually erode only a very thin section of the total material removed by denudation. Most of this material is, in fact, initially weathered and transported on the hillslopes adjacent to the rivers. The river itself acts mainly as the final method of transportation. The processes of hillslope transportation are therefore of enormous importance in the development of a landscape.

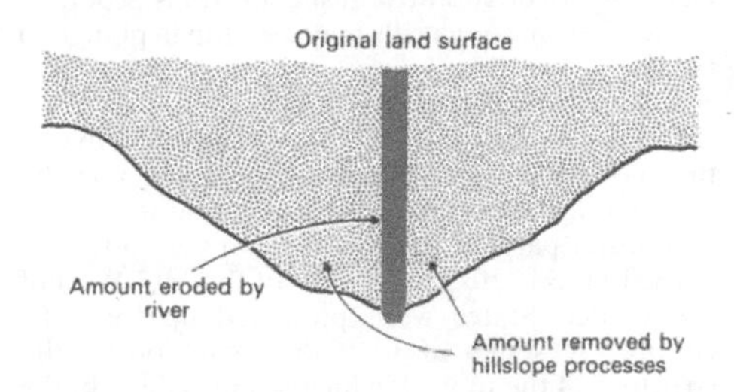

Figure 19 Cross-section of a river valley showing the amount of material removed by hillslope processes as opposed to that removed by the river.

Photograph 3 The products of combined physical and chemical weathering—a mantle of rock fragments and clay particles.

1 *Transport by water*

Water moving beneath the surface as throughflow or ground-water flow is capable of transporting the dissolved products of weathering to the river. Although this movement is not very spectacular (see Photograph 8), it is of vital importance because nearly all rock types provide some soluble weathered material. In addition, this is a process which continues to operate all the time: in both dry and wet conditions. The amount of soluble material which can be transported does not depend upon the speed of the water movement but is controlled by the **saturation capacity** of the water. This capacity is the maximum amount of dissolved material which can be held by a given volume of water, and it varies mainly with the acidity of the water. In the case of most silicate minerals, throughflow and ground-water flow remove all the soluble products faster than they can be produced by weathering, so maximum capacity is rarely reached. In the case of carbonates (chalk and limestone), however, weathering is rather more rapid and saturation capacity is easily reached, with the result that no more soluble bicarbonate can be created by chemical reaction.

On the other hand, water which moves through the spaces between soil and rock particles is not normally capable of moving undissolved rock fragments, because the spaces are smaller than the fragments. The exception is the case of pipes in soil, which are usually large enough to permit the carrying of solid fragments.

Water flowing over the ground surface as overland flow does not carry much dissolved material, because it has not entered the weathering zone within the soil. In contrast, overland flow is capable of picking up undissolved fragments on the soil surface. The size and number of particles which can be carried increase with the speed of waterflow. In part, the speed or **velocity** of flow depends upon the angle of the slope over which the water is moving, but velocity also increases with the depth of flow, because water very close to the surface is slowed down by friction. Overland flow depth during a storm will increase downslope as more water is being added by rainfall (Fig. 20) and velocity increases downslope as a result. On a natural hillslope, there are many minor irregularities in the surface, with the result that overland flow does not form a complete sheet on the ground but soon becomes concentrated into small rills (Fig. 21). Being concentrated in this way, the depth of flow in a rill increases and so does the velocity. Overland flow is then able to pick up material in the rill.

Although overland flow would appear to be a most effective process of hillslope transportation,

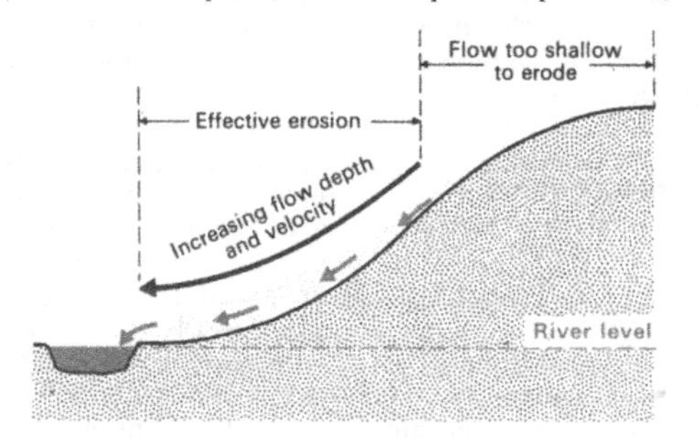

Figure 20 Transport and erosion by overland flow on hillslopes.

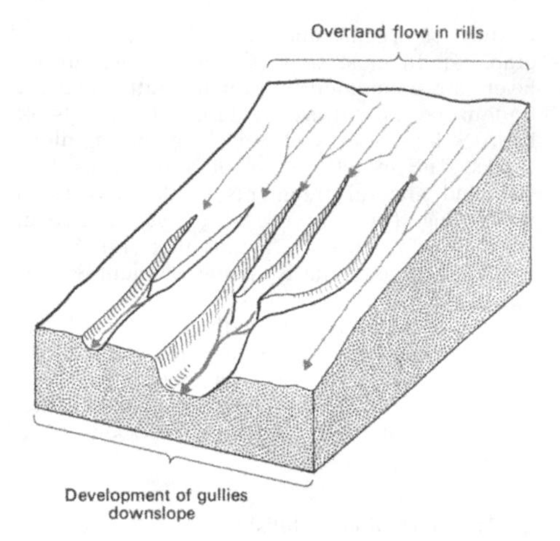

Figure 21 Soil erosion on hillslopes as a result of overland flow forming gullies.

there are grave doubts about its role in humid temperate environments today. In particular, it has already been observed (see page 14) that overland flow is something of a rarity on most hillslopes owing to high soil infiltration capacities and is at best restricted in frequent occurrence to slope base and hollow sites. Secondly, and perhaps more important when considering the ability of overland flow to remove and transport material, is average surface strength. Wherever vegetation is found, the root mat is normally capable of holding soil particles against the action of overland flow. Indeed, in Britain the climate is normally so damp that the topsoil retains strength even when ploughed, because soil structure is maintained. Certainly, overland flow in sheet form is generally ineffective, although rills may develop in ploughed fields.

It should be remembered, however, that overland flow might well have been more effective in the past and even very slow processes may be effective if they operate for long periods.

In other parts of the world it is a very different story. In the 1920s, large areas of the Mid-West of the United States were ploughed up for agriculture. A series of very dry years broke the structure of the topsoil, which fell into dust. In the heavy rains which eventually followed, overland flow rapidly developed first small rills and then huge gullies, leading to massive **soil erosion** and the

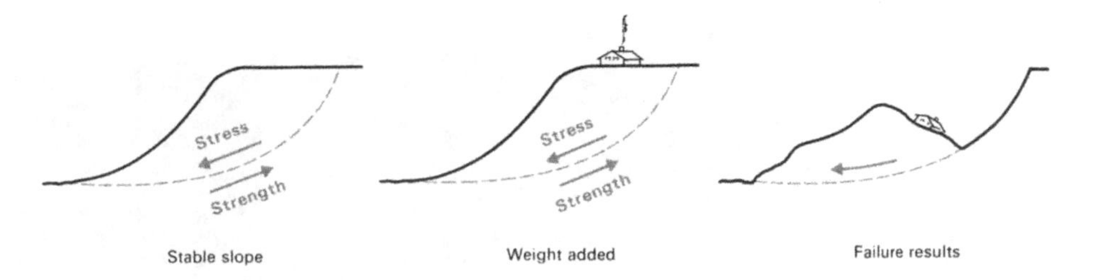

Figure 22 A hillslope may collapse if the weight of material being supported is increased.

total destruction of farming areas. An experiment carried out in the state of Mississippi at the time showed that while overland flow was removing 0·025 tonnes of soil from an acre of oak forest in one year, the same rainfall was removing 96·5 tonnes from every acre of land planted with cotton bushes.

2 *Transport by mass movement*
Weathered material can move down hillslopes under the influence of gravity and without the assistance of running water. This is called **mass movement**.

Gravity exerts a constant pull on all earth material (**stress**). Weathered material on a slope would immediately slide down to the river were it not for the fact that it possesses a certain strength. This strength is due to a number of material properties. First, all particles have rough surfaces, which results in **friction** between touching particles. Secondly, the shape of the particles may cause them to **interlock** rather as the pieces of a jigsaw puzzle interlock. Finally, the clay particles of weathered material attract each other by electric forces which produce **cohesion** between particles.

Mass movement occurs when the stresses acting on the material exceed the strength of the material. Usually this movement occurs slowly with stress and material strength nearly equal. Sometimes, however, mass movement occurs suddenly. Sudden movement may be the result of an increase in stress due to adding more weight to the material: building a house on the top of a slope, for example (Fig. 22). Alternatively, the strength of the material may be suddenly decreased by a rise in water content which forces the particles apart.

(*a*) *Rockfall*. In highland areas of Britain, near-vertical cliffs are fairly common. When rock fragments on the cliff are weathered by the action of freezing water or related processes, they will probably fall under the pull of gravity because there is nothing to bind the fragment of the cliff. This process is known as **rockfall**. At the base of the cliff, the fallen rocks will accumulate as a **scree** slope (Photograph 4). A particularly spectacular example can be found along the side of Wastwater in the Lake District.

Scree slopes are usually remarkably straight in section and in any one area tend to develop a common slope angle. This **angle of rest** seems to depend largely on the shape of the fragments (Fig. 23), which controls the angle at which the fragments interlock without falling. Within any one

Photograph 4 Rockfall and scree slope developments on a cliff.

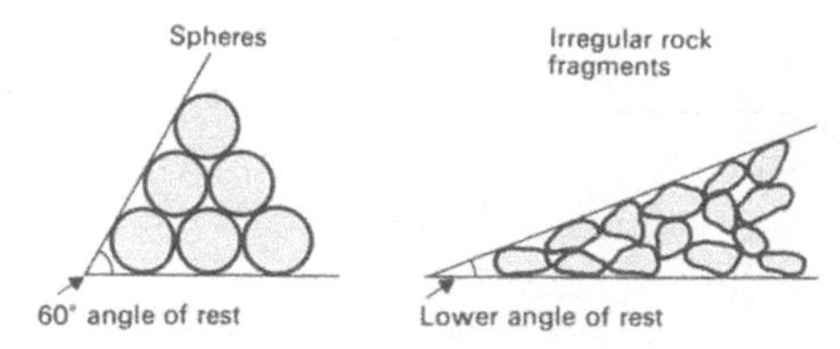

Figure 23 The angle of rest of loose particles depends upon the shape of the particles.

scree slope, it is possible to find an increase in the size of fragments down the slope (Fig. 24). Large blocks landing at the top of the slope can easily roll over the small fragments, whereas a small fragment will be trapped in the available spaces.

(*b*) *Landslides.* **Landslides** are the sudden movement of large masses of material rather than the individual blocks of rockfall. Usually this means that a large mass of soil on a slope gives way, but landsliding can also include *unweathered* rock if the forces involved are sufficiently powerful. In this case, landslides are more properly described as agents of *erosion* rather than simply transportation. Rockfall occurs more or less continuously but landslides occur mainly as the result of special conditions. In nature, these conditions are found most commonly where the slope is being undercut by a river or the sea, the result of which is to increase the slope angle and to make gravitational stress more effective. Small landslides can be found on the banks of many rivers. Along the Dorset coast (at Cain's Folly, Fig. 26, for example) are a series of enormous landslides. In this case, the sea has undercut the rocks, making them unstable, and the final collapse has usually come during a period of wet weather when the water content of the material has increased sufficiently to reduce its strength by reducing the friction and cohesion between particles. These landslides (Photographs 5 and 6) look particularly spectacular because the

Photograph 5 The development of tension cracks at the back of a large rotational landslide producing block sliding.

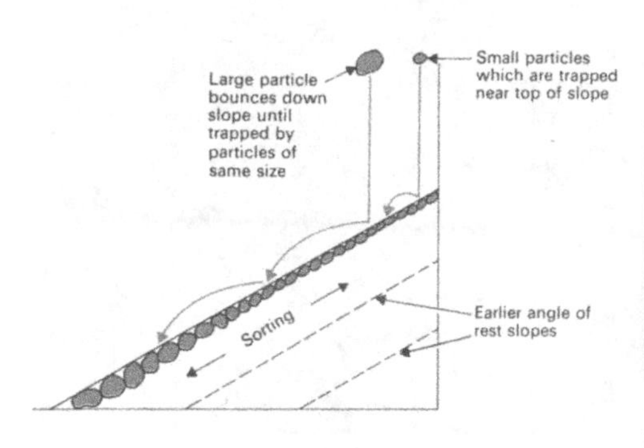

Figure 24 Larger particles travel farther down a scree slope.

Photograph 6 A large rotational landslide (Stonebarrow, Dorset) with a war-time radar station, formerly at the cliff-top, now on the reverse slope of the rotated block.

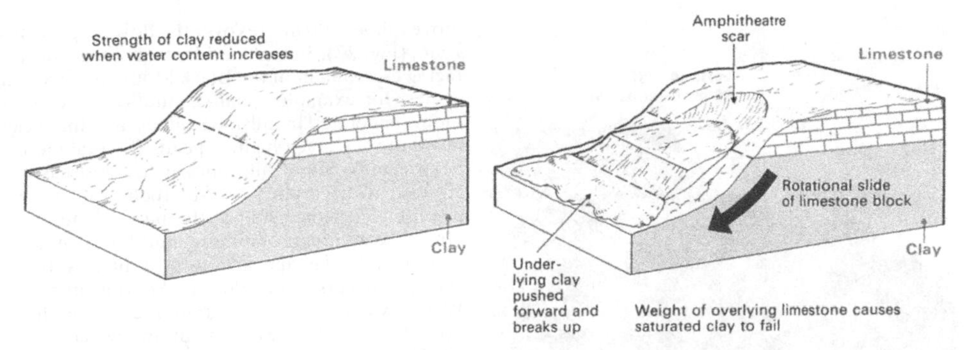

Figure 25 A rotational landslide caused by the failure of clay beneath a strong capping rock which does not break up during movement.

failure has taken place in clay rocks which underlie a stronger cap of sandstone or chalk. When the slide occurs, the solid capping rock does not break up entirely, and the rock mass often rotates as a single block **rotational slide**, leaving an amphitheatre-shaped scar in the hillside (Fig. 25). It is interesting to note that in Dorset the remains of old landslides can be found in similar situations inland (Fig. 26). It would appear that these were active in the past when the climate was somewhat harsher (see Chapter 6). Landslides still occur on normal slopes today when man interferes with the hillslope. The most obvious examples are to be found along our older motorways where the cuttings have been made too steep for stability. Landslides are also common in areas of the world subject to earthquake activity.

(*c*) *Mudflows.* Both rockfall and landsliding are the result of a sudden failure of material strength. Under some conditions, mass movement takes place gradually. If water accumulates in weathered clay, for example, the material eventually starts to flow downslope as a **mudflow**. These mudflows usually move slowly and are most active in winter, when rainfall is highest. Although rare on

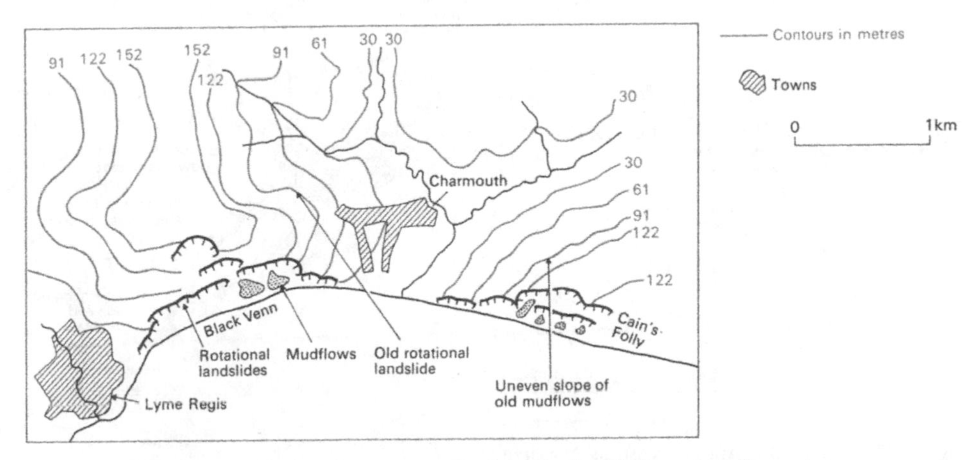

Figure 26 Part of the Dorset coast, showing active landsliding and mudflows where slopes are undercut by the sea and inactive sites farther inland.

Photograph 7 A small mudflow emerging from the bowl of a rotational landslide.

vegetated slopes, mudflows are common on the coast (Photograph 7) and are often associated with landslides, because the reversed slope of a rotational slide provides an ideal place for water to accumulate along with clay particles. To the east of Lyme Regis, on the Dorset coast, a huge area of the cliff is moving into the sea in the Black Venn mudflow. Each winter a new flow develops and moves about 100 metres over the flows of previous years (Fig. 27). In highland Britain, on the north-facing escarpment of the Black Mountain in South Wales, for example, smaller mudflows are commonly generated in hillslope gullies, and these can move at very much higher speeds (several metres per second). Some hillslopes in these areas are covered with parallel lines of rocks which have been left on either side of the moving mudflow. Under exceptional conditions, mudflows can cause considerable damage and loss of life. A tragic example occurred in 1966 at Aberfan in South Wales, when a water-saturated coal-slag heap turned into a massive mudflow which overwhelmed the village school.

(*d*) *Soil creep*. Although mudflows are rare on normal slopes, a slower form of the same process takes place almost everywhere. **Soil creep** is the very slow movement of soil down a hillslope. The precise nature of the movement is not really understood, but anything which causes the particles to move slightly, wetting and drying or heating and cooling of particles, plant activity or just animal

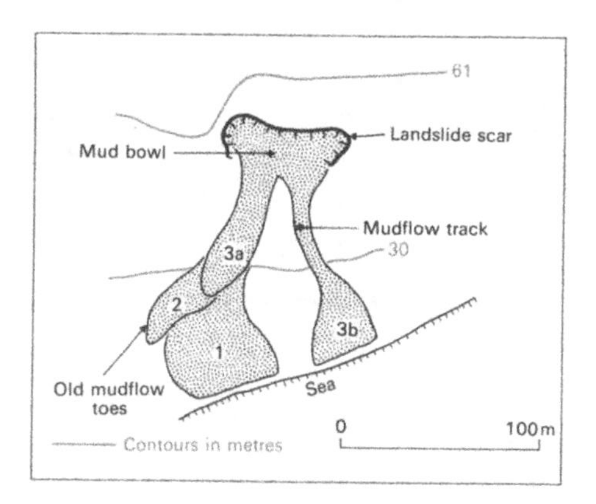

Figure 27 Part of the Antrim coast in Northern Ireland, showing mudflows which have occurred in successive years.

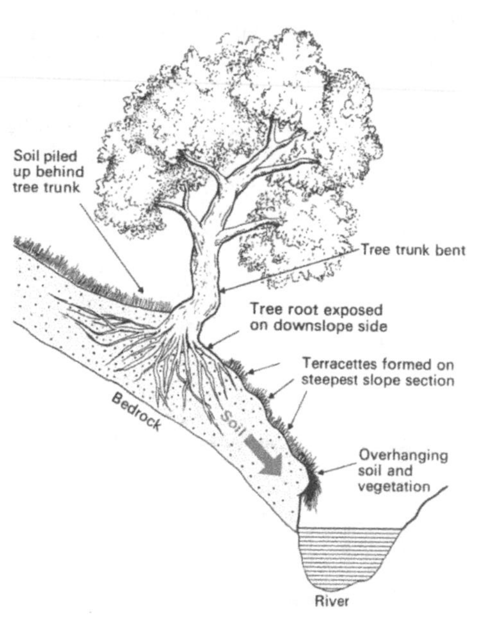

Figure 28 The features resulting from soil creep on a slope.

grazing, will allow gravity to pull the particles slightly farther downslope each time. Soil creep is very difficult to measure because it rarely goes faster than 1 mm/yr. Nevertheless, the effects of creep can be seen clearly in the bending of trees and the creation of **terracettes** on steeper slopes which are small breaks in the surface level (Fig. 28). Terracettes probably indicate that a certain amount of the 'creep' motion takes the form of minor slipping or fracturing.

Despite the fact that soil creep is very slow, it takes place all the time and on almost all hillslopes (see Photograph 8). It is therefore an extremely important process of hillslope transportation in Britain and may be compared with throughflow and overland flow in total effect.

3 *Transport process and hillslope shape*

The subject of hillslope shape is very complex. At the simplest level, hillslope shape appears to be the product of differing rates of weathering and transportation at various points on the slope. There are, however, three main problems in trying to understand hillslope shape. First, it is not always understood why a particular process should lead to a particular shape. Second, on any one hillslope, there will be a number of weathering and transport processes in operation. Third, the shape of the hillslope will eventually start to influence the type of weathering and transport processes. The complications of the problem are such that the following discussion is a deliberate simplification of the actual situation as it is understood.

On any hillslope, there is a relationship between the rate of weathering and the rate of transportation. On the one hand, transport processes can only carry away as much material as has been weathered. On the other hand, if the transport processes do not carry away all the weathered material, the depth of soil will increase, with the result that soil water acids are used up in chemical reactions before they reach the bottom of the soil and the rate of weathering at bedrock will decrease. This relationship permits consideration of two different situations. In one situation, the rate of transport is slow in comparison with the potential rate of weathering. This is the **transport-limited** case, where the transport rate holds back the rate of weathering. In the other situation, transport is capable of removing all the weathered material and is restricted by the rate at which weathering takes place. This is the **weathering-limited** case.

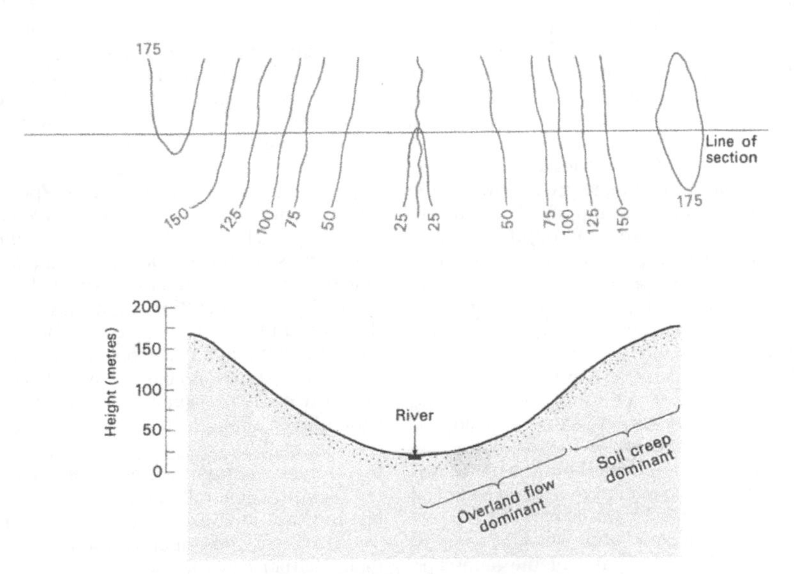

Figure 29 A map and cross-section of a valley with convex–concave slopes in an area where erosion is limited by the rate of hillslope transport.

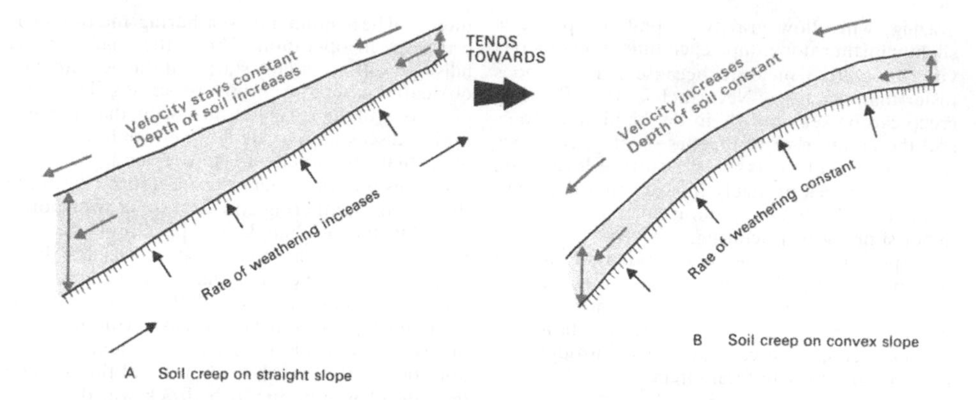

Figure 30 Soil creep on (A) a straight slope will usually result in faster weathering at the slope top, which must produce (B) a convex slope where weathering is the same over the whole slope.

(*a*) *Transport-limited slopes.* Over much of lowland Britain, the rocks are fairly weak sedimentary rocks which can weather quite rapidly. Weathering is limited by the rate of transport, which is fairly slow because the slopes are usually gentle and the rainfall low.

The most common hillslope shape in lowland Britain is convex–concave (Fig. 29). The upper, convex, part of the slope seems to be dominated by soil creep. Since more weathered material is added progressively downslope, it follows that either soil depth or velocity of soil creep must increase in a downslope direction. On a straight slope, creep velocity is not likely to change greatly and so the additional weathered material available downslope will produce a greater soil depth (Fig. 30). However, since soil depth controls weathering rate, the effect of increasing downslope soil depth is to produce a convex slope, as weathering would proceed more rapidly beneath the shallower soil at the top of the slope. Once a convex slope is established, the velocity of soil creep increases downslope, leading to a more constant soil depth and more uniform weathering rate. A convex slope is therefore self-perpetuating.

The argument is complicated but the analysis is probably correct, since it leads to a finely balanced equilibrium between the various parts of the system. It is this equilibrium state which appears to exist so commonly in the parts of the landscape that are not undergoing rapid change.

The concave section of the slope is probably produced by overland flow. The same basic shape is found in the long profile of rivers (see below) when the level of the sea is fixed. In the case of a hillslope, the base level is effectively 'fixed' by the river. Increasing downslope discharge on the lower part of the slope, due to incoming rainfall, is compensated in the decreasing gradient of slope. This fine relationship between discharge and gradient seems to exist because it leads to the most uniform pattern of energy expenditure on the slope (energy expenditure in terms of water movement being here represented by discharge times gradient).

As far as Britain is concerned, concave slopes appear to be common despite the points raised earlier about the general ineffectiveness of overland flow. This contradiction may imply that cave slopes are relics of an earlier age when overland flow was more effective owing to heavier rainfall or lighter vegetation, or it may simply mean that overland flow does leave an impression on the landscape although it is irregular in operation.

On some slopes, no basal concavity is apparent, but these cases are usually related to a river which is rapidly eroding downwards (see page 35).

(*b*) *Weathering-limited slopes.* In upland Britain, the rocks are generally more resistant to weathering and the rainfall is usually heavier. Transport processes are consequently often more efficient than weathering processes.

It is thought that this situation leads to the formation of long straight sections (Fig. 31) on

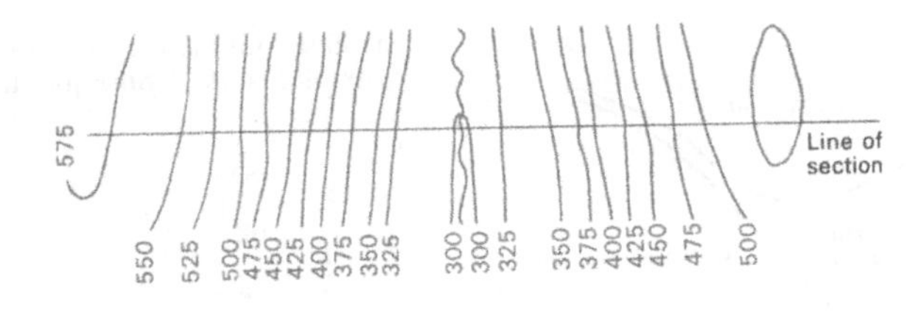

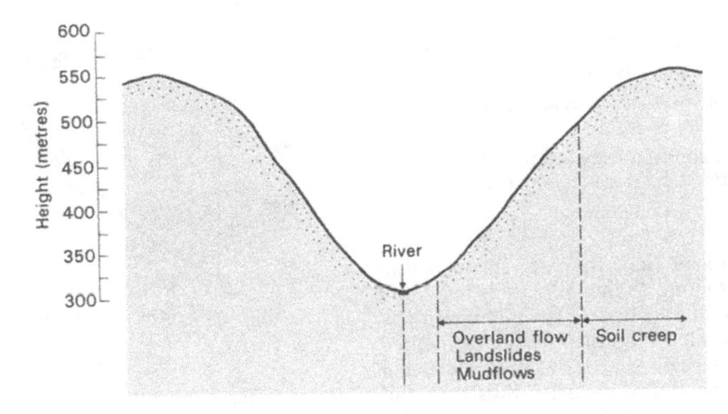

Figure 31 A map and cross-section of a valley with slopes which have a straight section controlled by the rate of rock weathering.

upland hillslopes. The slope base is still linked to the river and is therefore a transport-limited concavity, but the mid-slope is an environment where rapid transport by a variety of processes including the mass movement of soil layers and some types of overland flow takes place. In so far as the transport processes potentially operate more quickly than the weathering processes, a straight slope develops. The extreme example of the weathering-limited slope is the vertical cliff, where transport by rockfall is immediate and total, once weathering is complete.

Chemical weathering on limestones tends to produce a weathering-limited slope, since dissolved products are immediately transported by water. On other rock types, chemical weathering may be linked to soil depth and therefore be part of a transport-limited slope (see page 31).

D. Transportation, erosion and deposition in rivers

1 *Transport and erosion processes*

A river is constantly carrying out two functions. First, it carries away material produced in the denudation of the surrounding hillslopes. Second, it uses that material to cause the erosion of its own bed.

The **dissolved load** which the river carries is the product of chemical weathering on the bed of the river and on the adjacent slopes. Dissolved material from the slopes is brought into the river by throughflow and ground-water flow. A river is able to carry all the dissolved material supplied so long as the water is not actually saturated with a particular chemical. Saturation is not very common, and is most likely to occur at very low flow, when the chemical concentration increases.

The **particle load** of a river changes with the velocity of river flow. At low flow, the velocity decreases and little of the particle load is actually carried. During a storm, however, velocity increases as discharge rises and the size of particle which can be carried also increases (Fig. 32). The exception to this very simple rule is that small clay particles often stick together, with the result that silt-size particles are usually picked up first. Particles actually being carried by the river are referred

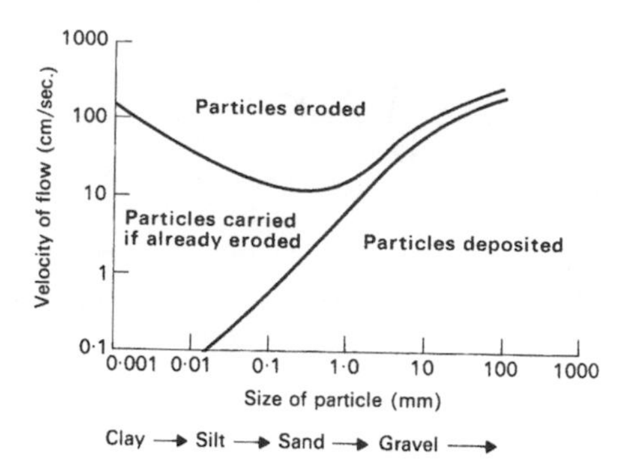

Figure 32 The relationship between the size of particle which can be carried by a river and the velocity of waterflow.

to as **suspended load** since they are suspended in the water only so long as the velocity is high, and they drop back to the bed again when velocity decreases. Suspended load drops back to the bed of the river in the reverse order to that in which it was picked up. Sand-size particles thus drop out before silt-size particles. In this case, however, clay particles are the last to drop and can be carried by a velocity much lower than that which was required to pick them up in the first place.

Some particles on the river bed are too large to be picked up properly, but these may be rolled along by the force of water as **bed load**. Strictly speaking, of course, it is impossible to distinguish between suspended load and bed load, because particles which are being rolled along the river bed on one day, when velocity is low, may well be in suspension on the following day, when velocity is high. The situation is further complicated by **saltation**. Some particles, too large to form suspended sediment under the prevailing conditions, may be dislodged into the current of water by other particles moving on the river bed. A dislodged particle may be swept along for a short distance before settling back to the bed, dislodging further particles as it does so.

A river obviously transports most of its particle load during periods of high discharge and therefore high velocity: that is, during flood events. Both the size and the number of particles increase with velocity. The result is that the *concentration* of particles in a volume of river water increases with velocity (Fig. 33A). At the same time, of course, not only will the concentration of particles increase, but the total volume of water (discharge) also increases. Consequently, the total particle

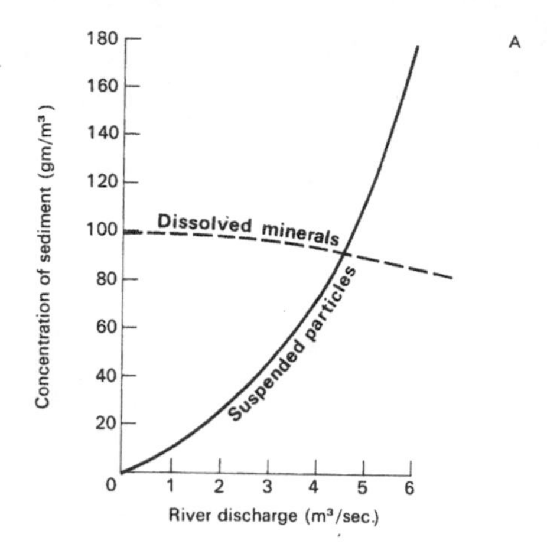

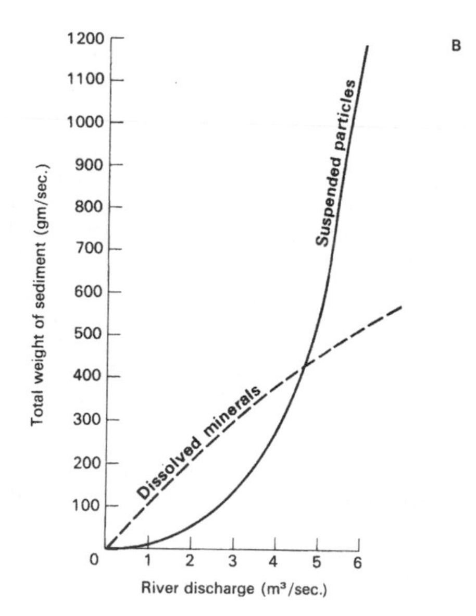

Figure 33 (A) Changes in the *concentration* of chemicals and particles carried by a river as the river discharge increases. (B) Changes in the *total weight* of chemicals and particles carried by a river as the river discharge increases.

load carried by a river shows a spectacular rise with increasing discharge (Fig. 33B).

The effect of changing discharge on the dissolved load is slightly different. First of all, an increase in discharge means that some water is entering the river as overland flow. Since overland flow carries little dissolved load, the concentration of solutes in the river will fall because of dilution. The dilution effect is counterbalanced by increasing discharge: each unit of water carries less dissolved material but there are more units. The overall effect is a gradual increase in total dissolved load with increasing discharge (Fig. 33B).

In addition to its task of material transportation, a river will also cause the breakdown of rock exposed within its channel by **erosion**. Erosion of the river channel may proceed partly by chemical action (**corrosion**) but physical processes are usually more in evidence. Of particular importance are:

(a) **Hydraulic action**. The combined effect of pressure and drag by moving water may cause the breakdown of rock. It seems likely that hydraulic processes are most effective where the rock is already well-fractured or jointed.
(b) **Abrasion**. As indicated above, rivers carry solid particles in suspension and on the stream bed. The effect of moving those particles over the surface of exposed rock is to cause abrasion (sometimes referred to as **corrasion**) of the rock surface. Abrasion rates obviously vary with the size, speed and quantity of particles involved and is hence most effective during flood conditions.

2 *Downstream changes in the river channel*

The discharge of a river normally increases continuously from its headwater to the sea as water is added to the channel by overland flow, throughflow, ground-water flow and tributary rivers. This increase in discharge causes corresponding changes in river processes (erosion, transport and deposition) and in the shape of the river channel.

In terms of channel shape, increasing discharge causes the width and depth of the channel to increase consistently downstream. Obviously, there are many irregularities in this pattern. On a small scale, any stretch of river shows a repeated sequence of relatively deep **pools** and intervening shallow sections (**riffles**) which seem to be related to the way in which water moves within the channel (see below). On a larger scale, bands of resistant

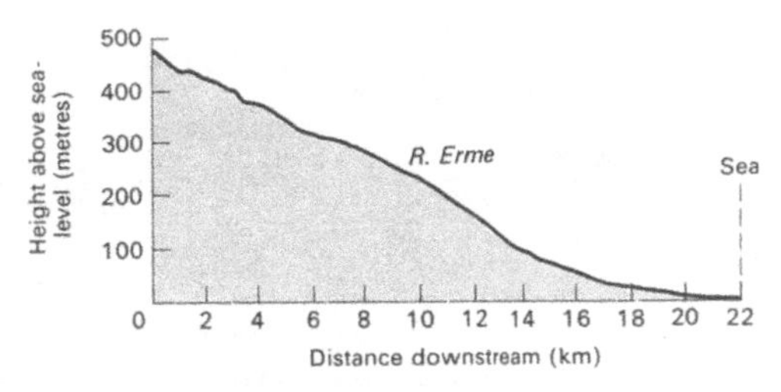

Figure 34 The long profile of the River Erme in South Devon from its source to the sea.

rock crossed by the channel may give rise to narrow but deep sections of channel (see page 43).

Similarly, the gradient of the channel generally seems to decrease downstream. Figure 34 shows a section drawn along the channel of the River Erme which has a distinctly concave lower portion. This drawn section is called the **long profile** of the river, and most rivers tend to have steep headwater and gentle lower reaches. The headwater region is not only steep, but is also often very irregular, with the channel gradient frequently interrupted by waterfalls or rapids. Such irregularities become far less marked downstream, although variations in rock resistance may still show, with the more resistant beds giving rise to steeper sections. The basically concave shape of the profile is considered to be an 'equal-energy' pattern. For any section of the channel, the river is turning the potential energy of height into the kinetic energy of water movement. Since velocity (more or less proportional to kinetic energy in this case) does not increase much downstream, it follows that the river is losing energy through the friction of water against the bed and sides of the channel. The energy loss for any section of river is equal to the volume of water (discharge) in that section multiplied by the gradient (that is the change in potential energy) for that section. In nature, the most probable situation is that energy loss for each section of river will be constant. If this is so, since discharge increases downstream, gradient will correspondingly decrease.

The headwater regions are usually the areas of maximum vertical erosion by the river. Since the river is ultimately eroding towards sea-level, it follows that the rate of downcutting is likely to be greatest in the highest portions of the river system. The results of physical erosion may be seen in the headwater channel as smooth sections of bedrock

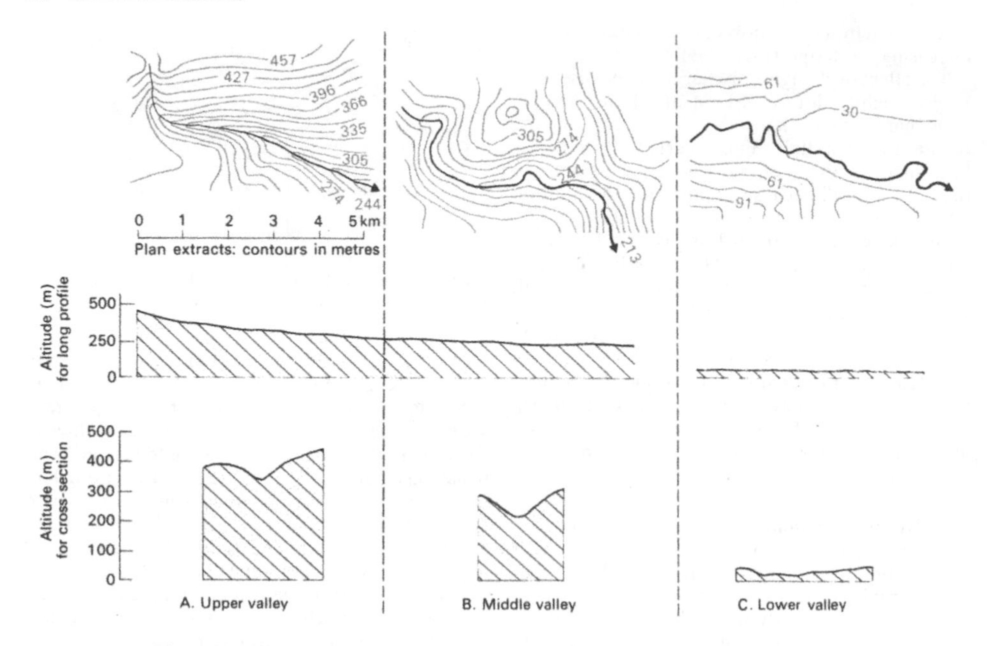

Figure 35 Three map extracts and corresponding long profiles taken from different sections of the River Exe in Devon. As long profile gradient decreases, the river becomes increasingly winding and the valley slopes become gentler.

in the channel, **potholes** created in the bed of the channel by the swirling action of bedload, or **plunge pools** beneath waterfalls. Farther downstream, the bedrock underlying the channel is often buried beneath a thick layer of bedload and may, in fact, never be subject to direct abrasion. Instead, in the lower reaches, rather more of the river's erosive power is concentrated upon the lateral erosion of the banks.

One other important downstream change which occurs as a result of the change from vertical to lateral erosion is the increasing sinuosity of the channel (i.e. in plan view, the channel becomes increasingly winding). The fairly straight channel of a river headwater (Fig. 35A) is gradually replaced by more and more exaggerated curves until **meanders** are formed which double back upon themselves completely. Generally, the straighter channel corresponds with steep sections of the long profile. Even here, however, the line of maximum water velocity seems to wander from side to side within the channel (Fig. 36) resulting in alternating shallow areas on either side of the channel. Under these conditions, it would appear that the steep channel gradient, by concentrating erosive power vertically, prevents the lateral swinging of the maximum velocity line from causing much bank erosion. On lower gradients, however, lateral erosion has more power and the natural flow pattern in the channel causes bank erosion and the eventual formation of a meander. As a meander becomes established, the curves of the channel cause the water to flow in a horizontal vortex and the line of maximum velocity is directed towards the outside of each bend (Fig. 36). Consequently, the meander form tends to develop further. There are, however, definite limits to this process. A major effect of a meander is to increase the distance that water has to travel down the river. As distance increases along a given river channel, the channel gradient decreases, since the height range is fixed. The critical limitations seem to be the relationship between discharge and gradient. Once a meander has produced the right gradient for the discharge at that particular point, there is no further exaggeration of the meander.

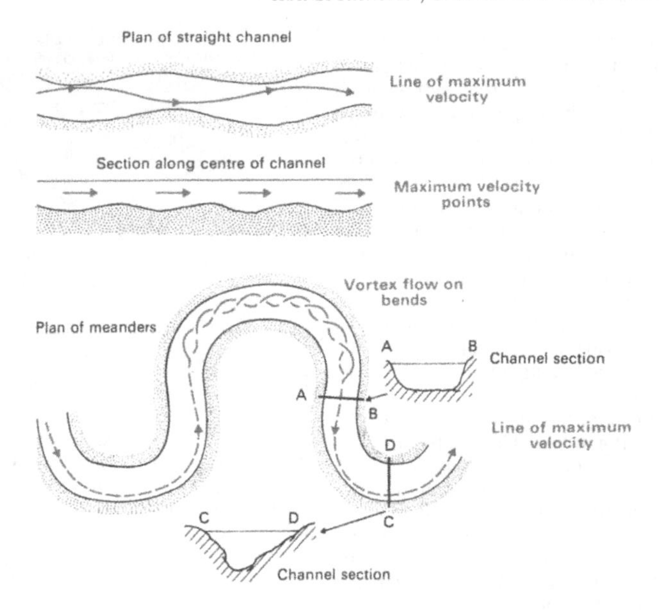

Figure 36 Comparison between the irregular flow of a straight river channel (*above*) and the steady vortex flow around the bends of a meander (*below*).

It is sometimes suggested that meanders occur because water flows more slowly over the lower gradients of the downstream section. In fact, this is a very false picture, because in most rivers the *velocity* of water flow actually stays more or less constant or even increases downstream. The reason for this apparent paradox is that although the channel gradient decreases downstream, the size of the channel increases. Since a good deal of the moving energy of water is absorbed by the friction of water against the bed of the channel, as the size of channel increases (Fig. 37), the proportion of the total flow affected by friction gets smaller. The decreasing effect of frictional drag compensates for the decreasing channel gradient and velocity stays the same.

Another frequent mistake is to imagine that erosion takes place in the headwater region of a river

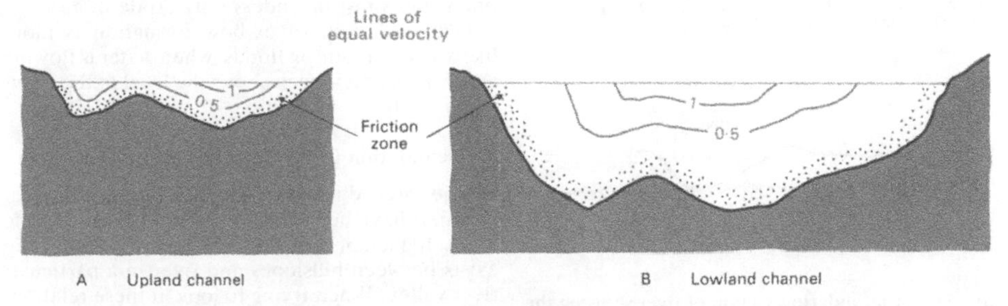

Figure 37 The contrast between (A) a small upland river channel and (B) a larger lowland channel which is less affected by edge friction and consequently has a similar average velocity, although flowing over a lower gradient.

and that the eroded material is then deposited in the lower reaches. It is important to remember first that, of the total sediment in the river system, most has originated from the hillslopes and not from the river channel. Furthermore, erosion of the river channel takes place at all points in the system but the relative importance of vertical and lateral erosion will vary. Inevitably, for most of the time, very little erosion or transportation (other than chemical load) takes place in the river system. During a flood, however, material is picked up at all points in the system, carried downstream some distance and then dropped again as the flood subsides. The main downstream variations occur because the amount of material picked up and later deposited increases downstream.

The most distinctive depositional feature of the river system is the **floodplain**. This is a flat area, on either side of the river channel, which is covered by water during a flood and which is created partly from the sediment dropped by the flooding river. The size of the floodplain increases downstream. The flat area upon which the floodplain deposits rest is actually created by meanders. Meanders are not fixed features of the landscape, since over time they 'migrate' down the valley by erosion on the downstream side of the bends. Generally, they retain the shape appropriate to the gradient on that section of channel (see above) as they move by depositing material on the upstream side of the bends (Fig. 38). After a period of time, migrating

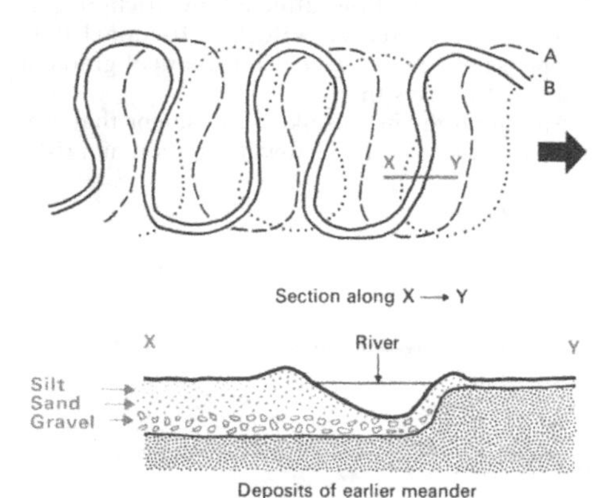

Figure 38 A meandering section of river showing the future positions of the meander (A and B). The section below indicates the nature of the deposits left behind as the channel shifts horizontally from X to Y.

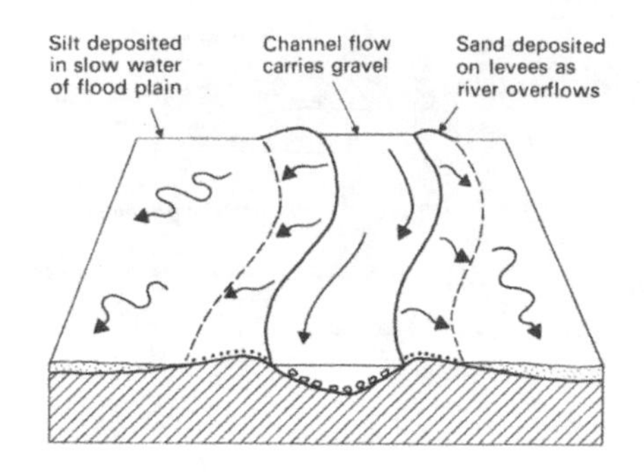

Figure 39 Deposits formed during a flood when a river overflows its banks.

meanders remove any obstructions in their way and leave behind a plain created by the deposition on the upstream side of bends. During a flood, when water cannot be contained in the channel, coarse sediment is dropped along the sides of the channel as water spills out on to the floodplain and loses velocity. Sometimes this material may actually form a distinct ridge on either side of the channel known as a **levee**. Occasionally, levees act as natural embankments maintaining the river at a level higher than the adjacent floodplain. Water which spills out on to the floodplain will carry only the finest particles (silt, etc.) and these will form a veneer of material on the surface of the floodplain after the water has receded (Fig. 39). One curious feature of the floodplain is the **ox-bow lake**. This is the remains of a former meander, now by-passed by the river, and left as a lake until silted up. It is possible that ox-bows might form as the result of opposite sides of meander eroding towards each other, but most meanders only erode in a downstream direction, so ox-bow formation is more likely to occur during floods, when water is flowing across a meander neck above the level of the channel (Fig. 40).

E. Denudation in the drainage basin as a whole

In the preceding sections, hillslopes and river channels have been considered largely as separate items. In fact, of course, a very precise relationship exists between hillslopes and river in a particular river valley. When trying to look at these relationships, the drainage basin, used for the analysis of water movement in Section A, will again prove to be the most useful unit of study.

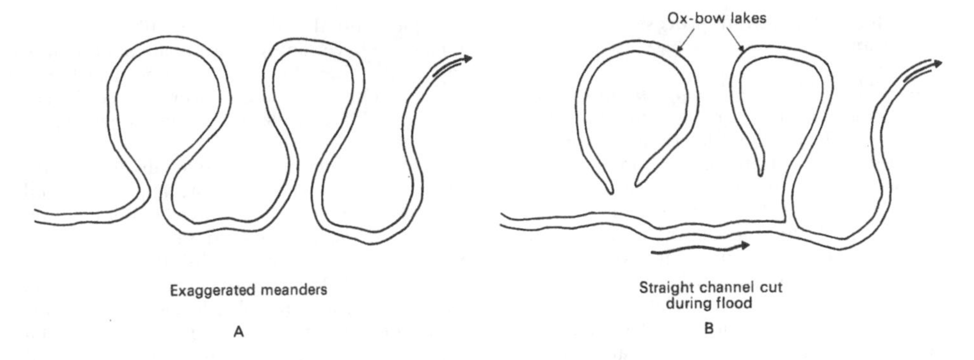

Figure 40 The formation of ox-bow lakes.

1 *Contrasts between drainage basins*
The products of denudation eventually leave the drainage basin via the river. The total volume of material removed from a drainage basin can be measured by the continuous recording of the dissolved and particle load of the river. This figure is usually expressed in cubic metres of rock per square kilometre of land per year.

Differences between drainage basins in terms of total material removed are largely a result of two main factors:

(a) Climate affects total water availability and therefore runoff volume. It also influences rates of weathering and possibly the rate of some transport processes which depend upon wetting and drying.
(b) Geology affects total relief and therefore both climate and average slope angle (which influences transportation rates). In addition, rock resistance tends to control weathering rates. Finally, rock type and therefore rock and soil permeability influence the time distribution of runoff and its consequent effectiveness.

Taken overall, these two factors help to explain major landscape contrasts between upland and lowland Britain. Within either one of these two areas, most of the contrasts between drainage basins are due to geological factors.

On a smaller scale vegetation or land use may have a considerable effect upon rates of erosion by

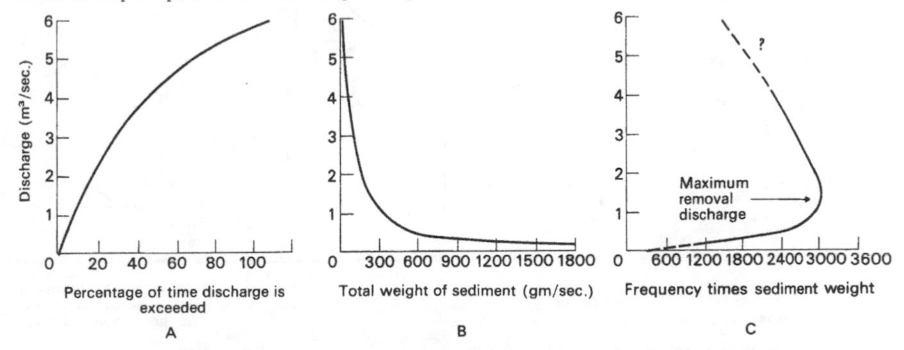

Figure 41 (A) The flow duration curve of a river combined with (B) the relationship between discharge and total sediment carried by the same river to produce (C) a graph which illustrates the total amount of sediment removed by each discharge level over a period of time.

their effect upon runoff processes and surface stabilisation.

2 *Variations within the drainage basin over time*
The sediment load in a river increases with discharge, and floods therefore remove large quantities of material. On the other hand, floods occur only occasionally, whereas low flows constantly affect the drainage basin. Which, then, is more important in the long term—the occasional flood discharge or the frequent low flows?

This question can be partly answered graphically. Figure 41A shows the relationship between discharge and the *total* load carried by a river and is thus derived from Figure 33, where suspended and dissolved load were kept separate. Figure 41B is the flow-duration curve for the river, originally presented in Figure 14 and showing the frequency with which different discharges occur. In Figure 41C, the two graphs are multiplied together to show the total amount of material removed by a particular discharge over a *long period of time*. The graph suggests that, although low flows occur frequently, they carry so little material that they are less effective than the less regular but higher flows. The big problem in this analysis is that there is very little information about the top end of the graph. Figure 41C implies that the very rare enormous flood is by virtue of its rarity, less effective than the intermediate flows. It is important to note, however, that this may not be true. In 1968, for example, a flood in the Mendip Hills of Somerset appears to have removed as much material in one day as *all* other flows have removed in the last 50 years. Obviously, the answers will not be clear until records are available for much longer periods of time.

In considering the foregoing discussion, it is important to remember that the analysis only indicates when the material is moved *out* of the drainage basin. The information says nothing about when initial weathering or hillslope transport may have occurred. Although river channel erosion and transportation are undoubtedly most effective during a flood, most weathering processes on hillslopes are unlikely to be much affected by individual storms. Of the hillslope transport processes, some, such as throughflow and soil creep, will not vary substantially during a storm, although overland flow and other mass movement phenomena are clearly linked to heavy rainfall events.

The overall picture would seem to be that during the long periods of 'normal' activity, weathering and hillslope transport produce material which is accumulated at the slope base since the river is not capable of its removal (Photograph 8). Only dissolved material, carried by throughflow and ground-water flow and very fine material in suspension will actually leave the drainage basin (Fig. 42A). During a flood, however, hillslope processes such as landsliding and mudflows may start to

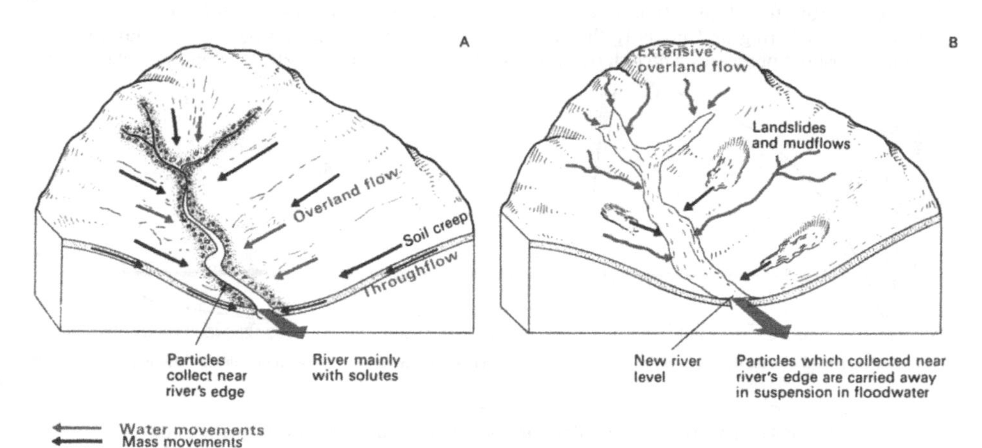

Figure 42 Under normal conditions (A), hillslope processes transport particles to the river which are not removed until a flood occurs (B).

Photograph 8
A small stream under normal flow conditions. Dissolved material from the slopes is brought to the stream by throughflow which emerges at the soil/bedrock boundary. Solid particles are carried by soil creep which gives rise to the overhanging banks. Solid particles are deposited alongside the stream which cannot transport them at low flow.

operate as the soil becomes saturated. Overland flow will become more extensive (Fig. 42B). Material from these processes will be carried to the river channel, where the swollen river will carry away both the fresh sediment and the material which has accumulated at the hillslope base since the last flood (Photograph 9).

The total denudation of a drainage basin is therefore partly achieved by slow continuous processes and partly by the more spectacular events of the occasional flood. The flood appears to be dominant, perhaps, since the bulk of sediment removal takes place at high discharge.

3 *Spatial variations within the drainage basin*
In a valley, the processes of denudation are concentrated towards the bottom. While soil creep and throughflow operate more or less equally at all

Photograph 9
Another section of the stream shown in Photograph 8 after a flood. The stream has first cut deeply into the loose material of its bed and has later spread a fan of bedload along either side of the incised channel.

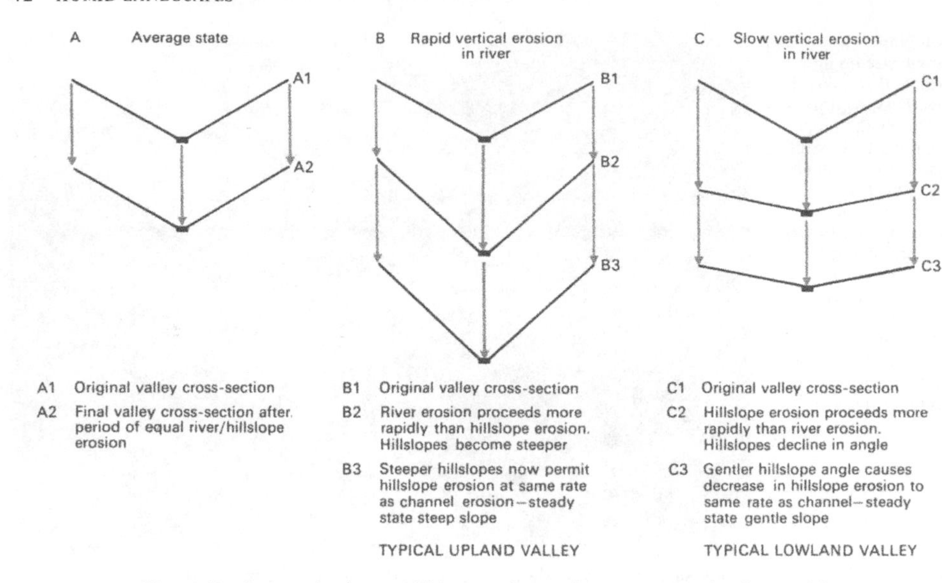

Figure 43 In an upland area, hillslope angle remains constant as the river rapidly erodes vertically. In a lowland area, hillslope angle declines over time because the river erodes vertically more slowly.

points on a slope, it will be fairly clear that overland flow occurs most frequently at the slope base and the processes requiring high rock and soil water content, such as landsliding and mudflows, will be similarly located. Maximum erosion, of course, takes place along the line of the river itself. The very existence of a valley then ensures its continuation over time.

The valley will not, however, simply deepen indefinitely, because there is a relationship between the river and its adjacent side slopes. In an upland area, for example, the river is normally eroding vertically downwards. If river erosion happens to proceed more rapidly than hillslope denudation in terms of average surface lowering, the hillslopes will become progressively steeper over time (Fig. 43). This increase in hillslope angle, however, will cause hillslope processes to operate more rapidly. Transport processes will strip the soil cover and the rate of bedrock weathering will consequently increase. Slope steepening will thus continue *only until* the increased rate of hillslope denudation matches the rate of river erosion. Thereafter the hillslope shape will remain constant over time. The same argument also works in reverse. If hillslope denudation is proceeding more rapidly than channel erosion (Fig. 43), material will accumulate at the slope base, thus reducing the rate of hillslope transport and slowing the rate of weathering beneath the soil until slope and channel are matched. The relationships which exist here are examples of the **dynamic equilibrium** between mutually interdependent parts of a balanced system. A change in any part of the system affects the other parts. The term **steady state** is applied to the condition of constant hillslope angle over time despite the continuing denudation of the landscape.

Within one river system, the rate of vertical erosion in the channel decreases downstream (see page 35). Since hillslope angle is related to rates of vertical channel erosion, it follows that average slope angle declines downstream and the valley thereby appears to widen out in cross-section. This range of steady state slopes within a single river system must not be confused with the age of the landscape. Taken overall, the landscape is progressively eroded towards sea-level during an erosional **cycle**. As relief declines during the cycle, so, too, do average slope angles. In one river system, how-

Photograph 10
A normally stable slope base which has slumped during a flood.

ever, hillslope variations downstream do not indicate that the lower part of the river system is 'older' than the upper portion.

4 *Spatial variations due to rock control*
The existence of more resistant rock bands within a drainage basin may cause aberrations in the simple pattern of long profile and valley cross-sections already noted. Within the river channel, a resistant rock band may cause a downstream increase in gradient (or even a waterfall), a downstream decrease in channel width and possibly a sudden straightening of the channel. On the adjacent hillslopes, a resistant band may produce a section of steeper slope and narrower valley. In extreme form, a river gorge may form across a hard rock band, a well-known example being the Avon Gorge on the River Avon downstream of Bristol, where the river flows through a section of Carboniferous Limestone.

The clearest indication of rock control, however, can be seen in the pattern of tributaries in a drainage basin. Where a river system develops upon a single rock type, a **dendritic** (tree-like) pattern of water courses forms (Fig. 44A). The existence of different types of rock will markedly affect this pattern, but it is impossible to make many useful generalisations in this direction, because there are so many possible geological structures. One common pattern, however, is provided by the sedimentary rocks of south-east England, which are tilted to produce a regular sequence of resistant and less-resistant strata (Fig. 44B). The resulting drainage pattern is a **trellis** pattern and consists of three basic river elements. **Consequent** streams are those which flow in the direction of the original geological structure. These then include the trunk stream and the tributaries flowing down the dip slope of the resistant rock bands. **Subsequent** streams are those developed at right angles along the line of the weaker rock bands. Erosion of subsequent valleys produces the typical landscape sequence of escarpment, dip slope and clay vale so characteristic of lowland Britain. Following the development of the subsequent valley, streams may be initiated which flow down the escarpments in the opposite direction to the original geological structure. These are **obsequent** streams.

Occasionally, trellis patterns may be found where an obsequent stream appears to have eroded back through the escarpment until it reaches and diverts the subsequent stream in that direction (Fig. 44). This process is known as **river capture.** It must, however, be emphasised that the suggested process does not have a clear theoretical basis. In particular, it is difficult to imagine how a river extends its catchment headwards, since it is the catchment which creates the river and not vice versa. A river cannot be maintained by a catchment below a certain size, but the phrase 'headwards erosion' implies that the river is eroding back into a progressively smaller area. It seems likely that many examples of 'river capture' are, in fact, examples of glacial diversion of drainage (see

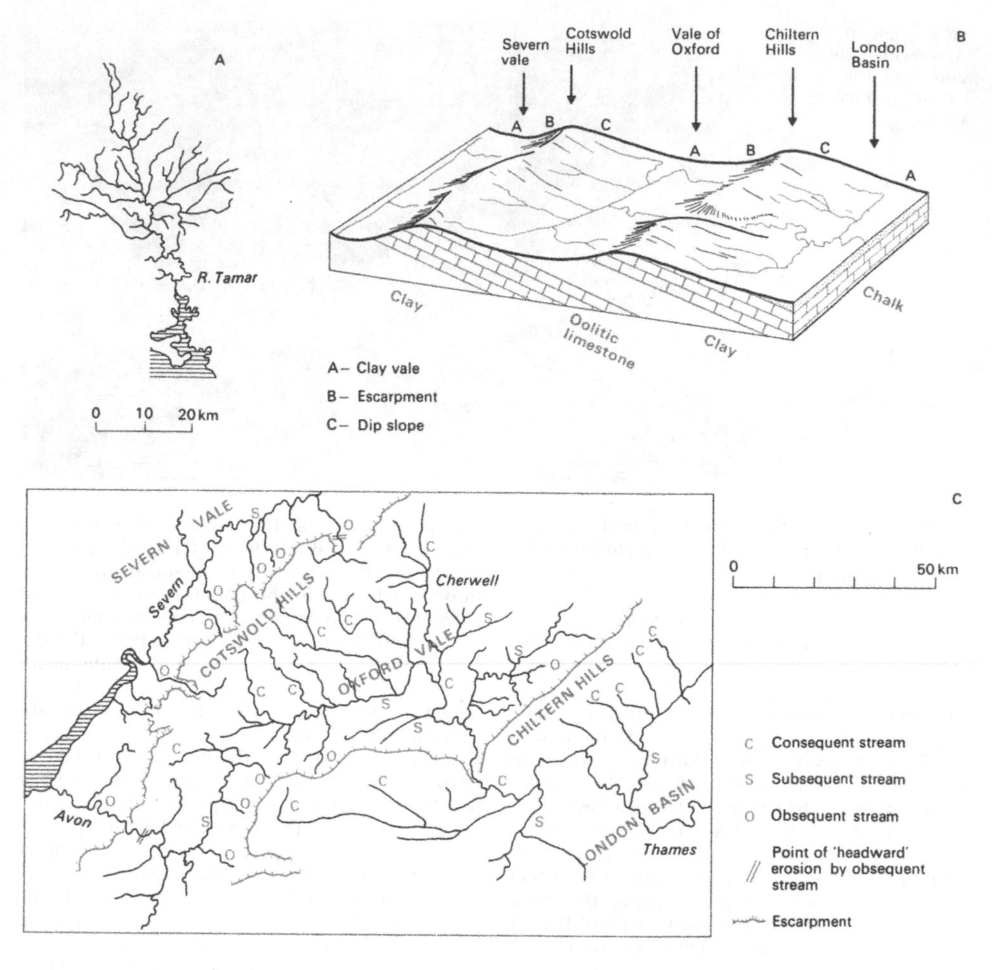

Figure 44 (A) The River Tamar on the Devon/Cornwall border as an example of a dendritic drainage pattern. (B) The River Thames and parts of the Severn and Bristol Avon systems. This is a very extended trellis drainage pattern, showing the main river elements and two examples of 'river capture' by obsequent streams eroding headwards. The block diagram shows the relationship between drainage elements and geology in the area.

page 87) or some other mechanism not yet understood.

F. The special case of limestone erosion

Variations in rock type produce differences in the landscape by their degree of resistance to weathering. Generally speaking, the type of processes operating do not vary substantially from one rock type to another, although the relative importance of different processes may vary. Limestone, however, is slightly different for two reasons. First, the

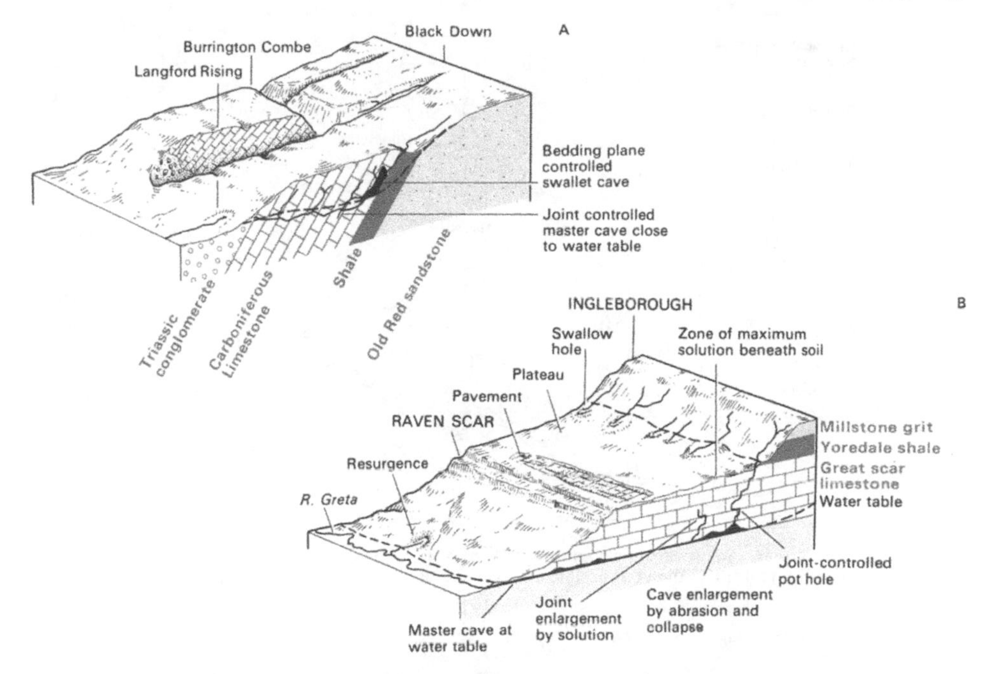

Figure 45 The surface and subsurface features of crystalline limestone shown diagrammatically for: (A) the northern flank of the Mendip Hills, Somerset; (B) the River Greta—Ingleborough area of Yorkshire.

process of limestone erosion is achieved by chemical weathering of a carbonate mineral which leaves very little insoluble residue to be transported on the surface. Second, limestone is permeable, with the result that much of the erosion, in fact, takes place beneath the surface.

The word 'limestone' actually includes a number of rock types. In Britain, the group includes the Cretaceous Chalk of south-east England, the Jurassic 'oolitic limestones' of the Cotswold Hills and the hard crystalline Carboniferous Limestone of the Pennines, South Wales and the Mendip Hills of Somerset. It is this latter limestone which gives rise to the most distinctive landscape.

1 *The movement of water in limestone*

In the case of Carboniferous Limestone, the rock itself is *not* permeable, because the calcium carbonate minerals of which it is made are in crystalline form, there being no spaces between the crystals. On the other hand, clearly marked breaks exist between each bed of rock (bedding plane), and each bed is also dissected by vertical joints which result from cracking in the rock. Water flowing through these routes gradually enlarges the spaces by chemical weathering, and the permeability of the rock increases. Eventually, the whole rock mass is dissected by many enlarged water routes, and the limestone is then capable of absorbing all the rainfall which normally occurs. At this point, all drainage takes place beneath the surface, with rainfall passing through the soil and percolating rapidly down to the ground-water zone. The water table in limestone is usually very flat and is controlled by the lowest outlet for water in the region. Because water moves through a number of large holes in the rock, rather than through the rock itself, much of the water for an area emerges at very large springs called **resurgences** (Fig. 45).

Normally, limestone areas are completely without surface rivers, except where a valley intersects

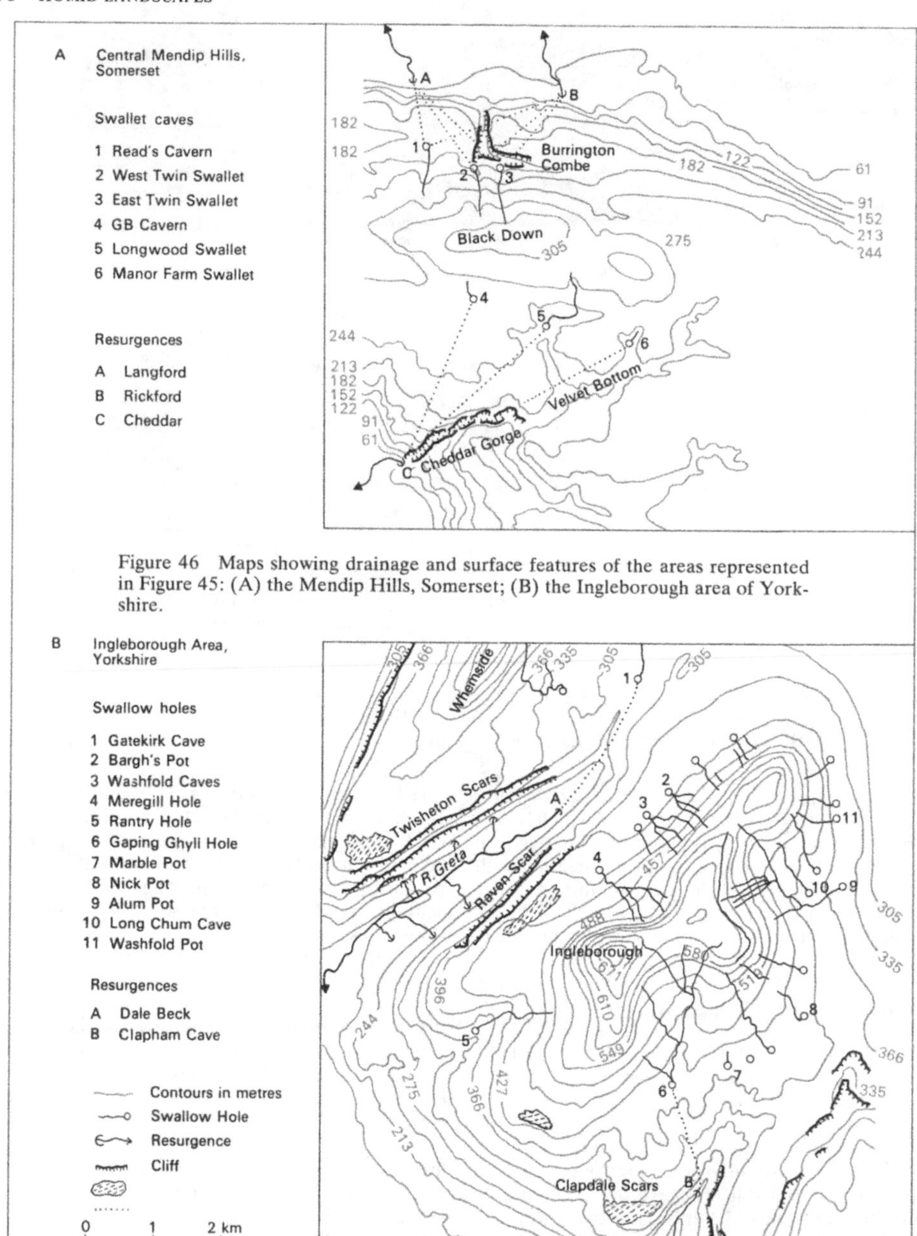

Figure 46 Maps showing drainage and surface features of the areas represented in Figure 45: (A) the Mendip Hills, Somerset; (B) the Ingleborough area of Yorkshire.

the water table. Under some conditions, however, rivers may originate on an adjacent area of impermeable rock and then flow on to limestone at some height above the water table (Fig. 45). Under these conditions, the river will **sink** into the limestone either through a number of holes or through one major opening (**swallow hole** or **swallet**). A river of this type will descend rapidly to the water table, where it may continue to follow a discrete joint or bedding plane to the resurgence. Whereas water percolating through the surface of the limestone takes several weeks to emerge at a resurgence, a disappearing river may continue to flow beneath the surface at surface river velocities. The streams sinking in the Burrington area of Mendip (Read's Cavern, East and West Twin Brook in Fig. 46A), for example, re-emerge at the resurgences of Langford and Rickford within twenty hours. Furthermore, the water of the various streams does not seem to mix at the water table as would happen in the ground-water zone of a normal permeable rock but the streams follow quite separate routes right to the resurgence. Water at the resurgence may therefore be either 'percolation water' which has entered the rock as rainfall and eventually reached the dispersed ground-water zone or 'swallet water' which has followed a separate route from swallow hole to resurgence.

2 *Erosion in limestone areas*

Erosion by limestone solution (see page 23) does not seem to take place equally throughout the limestone system. The major control on the solution process is the availability of carbon dioxide, which determines how much carbonic acid will be formed. Generally speaking, the highest concentrations of carbon dioxide seem to be either in the soil or in the joints of limestone close to the surface. It follows that most of the chemical weathering of limestone is carried out by percolating rainwater close to the surface. That water will, in fact, reach chemical saturation quite quickly and will not, therefore, be able to erode limestone farther beneath the surface.

It is rather strange that caves, always thought of as evidence of limestone solution, are often, in fact, places where little solution takes place. Generally, they are of two main types: those formed above the water table by percolating water (**vadose** caves) and those formed beneath the water table by ground-water flow (**phreatic** caves). Vadose caves close to the surface may develop into fairly open joint systems, but other vadose and phreatic caves generally will be small, since the rate of limestone solution at depth is going to be slow as available carbonic acid is used near the surface. The development of major cave systems tends to be restricted to two situations. First, in the vadose zone, major caves form along the routes followed by streams sinking into the limestone surface (**swallet caves**). The shape of the cave system varies with the geological structure: in the Ingleborough area of Yorkshire (Fig. 46B), the swallet caves such as Gaping Ghyll often start with a vertical drop of several hundred feet (a **pot hole**), while in the Black Down area of the Mendips, swallet caves such as Longwood Hole are steeply descending passages following tilted bedding planes. Second, major caves form at the water table behind resurgences where separate water routes appear (e.g. Cheddar Rising in the Mendips or Clapham Cave in Yorkshire). In both situations, however, only part of the cave development is due to solution. The swallet streams, for example, start on the surface, where the carbon dioxide concentration is low and all the chemical potential of the water for dissolving limestone is used up within a few metres of the cave entrance. It is very noticeable, for example, that water dripping from cave roofs deposits calcium carbonate as stalactites (in the ceiling) or stalagmites (on the cave floor) because the cave air has a lower carbon dioxide concentration than the narrow joints through which the water has percolated. While these caves must have been initiated by solution when they were still very restricted passages, much of their subsequent development is due either to abrasion of the cave floor by the sediment brought in by the swallet stream or by the collapse of the cave roof over the streamway.

It is not only beneath the surface that the distinctiveness of crystalline limestone can be seen. The surface of limestone areas has a number of features which occur because most of the drainage is underground. Where the percolating water of an area is concentrated into one vertical rock joint, the erosion of that joint may cause the collapse of the surrounding ground surface and the production of an enclosed depression (sometimes called a **doline** when occurring on a large scale). Usually more spectacular is the collapse depression formed at the entrance of a swallet cave. Occasionally circumstances occur which cause the drainage of a large area to flow inwards to one point, which develops as a large enclosed basin by solution and collapse. In Britain, it is rare to find many such basins more than a kilometre or so in diameter, but in other limestone areas of the world enormous basins may be found (for example, the **poljes** of Yugoslavia).

Underground drainage normally has the result of preserving any shape in the landscape, since there are no transport processes operating down the surface of hillslopes. Two features are particularly noticeable in that respect: plateaux and gorges. If a plateau surface exists in normal rock, it is soon obliterated as water and weathered material move over the edge of the plateau and gradually wear back its edge. On limestone, however, a plateau, once created, will be maintained as erosion operates equally upon all areas of the plateau, the weathered products being removed underground. In Yorkshire, for example, the hills of Ingleborough and Whernside (Fig. 46B) are surrounded by high-level limestone plateaux formed by horizontally laid limestone. In the Mendips, a plateau, created by denudation at a time of former high sea-level, is still preserved as solution continues to lower it.

Gorges seem to be common in limestone areas for a similar reason. On normal rock types, there is a link between river and hillslope because the river transports away the weathered debris from the slope (see page 33). An increase in river erosion causes the more rapid removal of slope material, an increase in hillslope weathering and an increase in total hillslope denudation rates to match the new rate of river erosion. A river flowing across limestone, however, creates a valley in which the hillslopes are eroded directly by the solution process, which leaves little residue for the river to transport. Increasing erosion by the river does not, therefore, increase the rate of limestone erosion on the hillslopes. Over a period of time, it is possible for a river to erode a very deep gorge (such as Gordale, near Malham in Yorkshire) the sides of which have scarcely retreated back from the channel cut by the

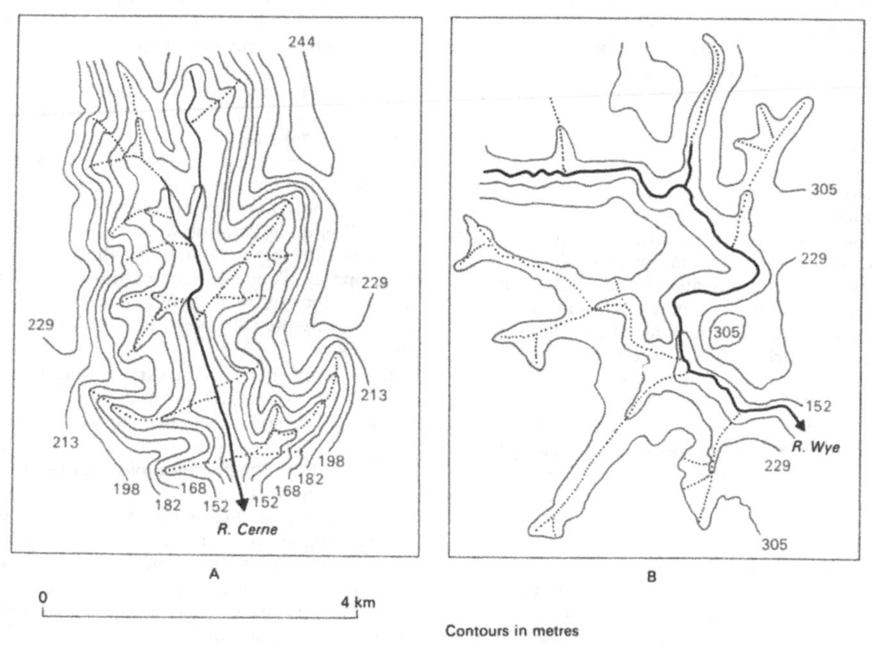

Figure 47 The pattern of dry valleys in (A) Chalk downs around Cerne Abbas, Dorset, and (B) Carboniferous Limestone in the valley of the River Wye, Derbyshire.

Photograph 11 The power of a river in flood is shown by the road embankment which was ripped apart during the Mendip flood of July 1968. The valley floor downstream of the embankment is completely covered by debris.

river. Vertical-sided gorges like this eventually favour the initiation of rockfall and scree development at the base of the cliffs, but this speeding up of the hillslope processes only occurs after the gorge has formed.

A further result of the general lack of insoluble residue from weathering is that limestone soils tend to be very thin, and bare rock outcrops are common throughout the landscape (the 'scars' of the Greta valley, Fig. 46B). Among the most spectacular of bare rock outcrops are the large areas of almost horizontal bare limestone called **limestone pavement** (Fig. 46B). Pavement takes its name from the dissection of the limestone into more or less rectangular blocks by lines of vertical joints. The blocks themselves are often highly sculptured by solution. It is thought that pavement is probably a relic from the Ice Age, when ice stripped the thin soil cover from such areas.

The extreme permeability of limestones also leads to the phenomenon of **dry valleys**, such as Velvet Bottom in the Mendip Hills. All areas of limestone (crystalline and oolitic) and chalk have networks of dry valleys. Limestone areas often show a fairly shallow network leading to a major dry gorge (e.g. Cheddar Gorge in the Mendip Hills). Chalk areas often show quite deeply incised dry valleys both in escarpments (**combes** such as Devil's Dyke near Brighton) and dip slopes (Cerne valley; Fig. 47A). A whole string of theories have been advanced to explain these different features: for example, the collapse of a cavern to form Cheddar Gorge; the gradual drop of a water table as an escarpment is eroded backwards; the existence of 'mud glaciers' during the Ice Age. It seems fairly clear that all these features have a common origin, and generally it is now assumed that dry valleys were eroded by normal rivers at a time before subsurface solution increased the permeability of the rock to its present level. It is likely that the basic valley network was created before the Ice Age and was reactivated during the Ice Age. In this case, it is thought that areas close to the ice sheets would be frozen in winter but in summer the surface layers would melt (leaving the subsoil frozen). Surface water, unable to penetrate the ground, would create rivers again in the dry valleys. It is noticeable that many gorges are formed in areas where sea-level was dropping rapidly during the Ice Age, and these may therefore be features of Ice Age rejuvenation (see Chapter 6) which were not part of the original dry valley network.

Chapter 3

Arid and Semiarid Landscapes

A. Water in an arid region

The word 'arid' is usually taken to mean very dry. Areas are arid, however, not simply because they receive very low rainfall, but also because the moisture balance between precipitation and evapotranspiration is negative: that is, annual evapotranspiration exceeds annual precipitation. On this basis, parts of the polar regions where rainfall is low are excluded, since evaporation is also low. Arid areas of the world then *approximate* the areas with less than 250 mm of rainfall per year. This includes (Fig. 2) the temperate deserts of Mongolia (the Gobi), Turkestan in Russia and Patagonia in Argentina, as well as the hot deserts of North Africa (the Sahara), South Africa (the Kalahari), Asia (Arabia, Iran and the Thar Desert of India), North America (the deserts of California), South America (the Atacama Desert) and finally Australia (for example, the Gibson and Simpson Deserts). Adjacent regions with less than 500 mm/yr and high evaporation rates may be classified as 'semiarid'.

In Chapter 2, processes were described which operate in humid areas of the world, and virtually all those processes were based upon the free availability of water. When we examine arid and semiarid landscapes, it must be remembered that fundamentally similar processes operate but that their relative importance is different, since water is not abundant. Very few processes in arid regions involve no water whatsoever, and contrasts between humid and arid landscapes tend to arise because of the irregularity of water availability and the general lack of surface vegetation in arid areas.

Because evaporation actually exceeds precipitation, permanent rivers do not originate within arid areas. On a world scale, several notable examples exist of rivers, such as the Nile or the Colorado, which rise in humid zones and flow through arid regions to the sea. Such rivers, of course, lose discharge as a result of either evaporation from the river surface or infiltration from the river bed into underlying rock, which is unlikely to have a high water table. Smaller permanent rivers flowing into deserts are likely to lose discharge until flow ceases. Typically, such rivers end in a lake (for example, the River Jordan ends in the Dead Sea) and thus form **basins of internal drainage**. Internal drainage systems are not controlled by sea-level and may therefore lie below sea-level. The Dead Sea lies at a level of approximately 400 m below the level of the Mediterranean. A further important feature of internal drainage systems is that they tend to be controlled to a great extent by geological structures. In the south-west of the United States, for example, the desert landscape was originally shaped into upstanding blocks and downthrown basins by faulting of the earth's crust. Rivers generated on the wetter upstanding blocks flow into the adjacent basins, where they evaporate. Under humid conditions, each basin would fill with water until overflow, when the basins would be united by a common river. Under arid conditions, however, each basin remains separate (Fig. 48).

In most arid regions, no permanent rivers exist. Instead, rivers flow for a short time after rainfall as **ephemeral** or **intermittent rivers**. A characteristic of desert hydrology is the extreme irregularity of rainfall. A figure of 50 mm/yr, for example, may disguise the fact that all 50 mm falls in two storms. In some areas precipitation is measured in terms of years with no appreciable rainfall. The lack of vegetation and frequently bare rock surface of the true desert may result in low infiltration rates (see page 13) with the consequent paradox that rainfall in arid areas is more likely to produce surface runoff (overland flow) than rainfall in humid climates. Rapid runoff of this type leads to **flash floods**, which may cover the desert surface within minutes.

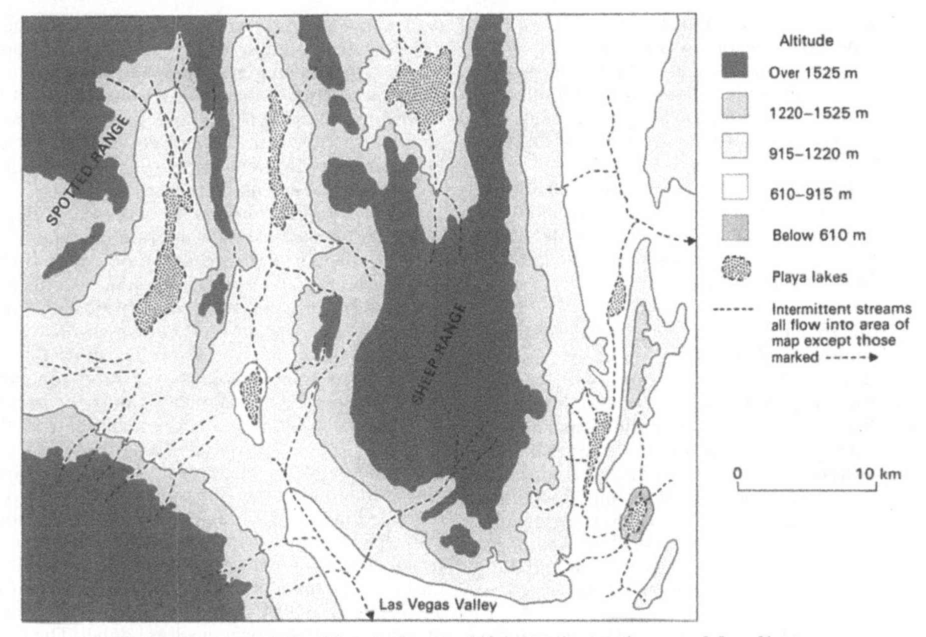

Figure 48 Part of the Nevada Desert, USA, to the north-west of Las Vegas, showing enclosed basin and mountain range landscape. Most of the drainage is intermittent and internal towards the centres of basins, where playa lakes can be found.

B. Weathering in arid regions

The general lack of moisture in deserts means that chemical weathering will be very much slower than in humid climates, and, indeed, for many years it was assumed that weathering in deserts must proceed entirely by physical means. It is now generally agreed, however, that chemical weathering (see page 22) does take place, although the rate at which it operates and its relative importance in total rock weathering are not clear.

On the other hand, the harsh desert environment makes it easy to imagine a number of physical weathering processes, although that does not, of course, guarantee their importance.

(a) Insolation weathering. Solar heating (insolation) in the day, followed by rapid night-time cooling in the absence of a cloud cover, lead to diurnal temperature ranges in deserts of up to 50 °C. This rapid change should set up stresses in near surface rocks as expansion and contraction occurs. In fact, it does appear that most rocks can withstand this kind of variation for long periods of time without cracking, and the process may not be as important as was formerly thought.

(b) Freeze–thaw action. Surprising as it might seem, desert areas often experience night temperatures below freezing point. Not only does the lack of cloud cover cause loss of heat by radiation, but also areas in the centre of desert basins tend to receive the 'drainage' of cold air from surrounding slopes. Parts of the Mojave Desert, California, may experience freezing temperatures on one-third of all nights in the year. It must be emphasised, however, that freeze–thaw requires not only low temperatures, but also the presence of water. Except for occasions after rainfall, therefore, freeze–thaw is unlikely to be important.

(c) Growth of salt crystals. In conditions of high evaporation, various salts accumulate near the earth's surface. It is thought that the growth of salt crystals, either during accumulation or when water is added after dehydration, may set up sufficient stresses to cause localised rock splitting.
(d) Changes in volume on wetting and drying. Some clay rocks expand when water is added. Desert environments provide periodic drying and wetting cycles which may cause cracking following expansion.

Although all the weathering processes mentioned above probably occur in desert areas at some time, none of them appears to be well developed. In part, this implies that weathering by other means may be as important, particularly chemical weathering, but also more general physical weathering such as unloading (see page 22). More to the point, perhaps, is the implication that weathering rates in deserts are slow. In general, desert surfaces show only a shallow weathering layer. Bare rock outcrops are common and many other areas have a **stone pavement** of coarse material.

One rather odd weathering product of the desert environment is the black or dull-red coating of isolated boulders and rock outcrops known as **desert varnish**. The varnish is actually a very thin layer rich in iron and manganese oxides. Originally assumed to be a surface deposit caused by evaporating water carrying the chemicals up to the surface, desert varnish is now more often thought to be a weathering product.

An important consequence of the general lack of weathered material at the surface is that further weathering and erosion often works upon bare rock surfaces. Geological weaknesses, either particular rock bands or joints and bedding planes, are therefore exploited selectively, and grotesque shapes may result at a local scale.

C. Hillslope and channel processes in arid landscapes

As might be expected, some types of mass movement are common transport processes on desert hillslopes. The generally slow rate of denudation and lack of weathering mantle results in many steep bare rock surfaces upon which rockfall takes place. Scree slopes transport most of the rockfall material, but there are also many slopes on which debris appears to be moving, although the slope is at less than the angle of repose. On such slopes debris is not moving by rolling and sliding. These slopes are **debris flows**, in which movement is by downslope creeping following repeated expansion and contraction with diurnal temperature cycles. Other mass movement forms such as landsliding are probably less common in deserts.

Surface runoff (overland flow) on desert slopes is quite possible during intense rainfall events (see page 50). The lack of vegetation on desert slopes means that surface runoff is capable of effecting considerable erosion. It is probably most effective when water is concentrated into small rills which cross the slope, but **sheetflow**, in which flow is spread across the whole slope, reputedly also occurs on very flat desert surfaces and may succeed in stripping layers of weathered material from the slopes.

Permanent river channels in deserts will obviously function in much the same way as the rivers described in Chapter 2. Ephemeral channels, however, have one or two peculiarities. Dry periods result in the gradual accumulation of material in the empty channel of the river. Flash flooding usually results in riverflow taking the form of a 'wave' of water travelling down the dry channel. The front of the wave seems to transport large quantities of accumulated material as bedload rather than suspended sediment. A new channel form is produced during the flow which subsides rapidly, leaving piles of bedload in the channel. The front of the wave dissipates rapidly downstream owing to infiltration into the channel bed once rainfall has ceased. The driest basin areas of deserts tend, therefore, to receive a much reduced riverflow which carries only the finer material to the centre of the basin.

D. Landforms produced by water-based processes

Generally slow rates of denudation in deserts mean that many of the major landforms of arid areas are largely geologically controlled. Despite the irregularity of rainfall, the major denudational landforms of deserts are still those produced by water-based processes.

Many deserts show roughly the same overall landscape components. In particular, enclosed basins surrounded by mountains are very common features (Fig. 49A), although the proportion of mountain and basin may vary enormously to the other extreme of an isolated block of upland sur-

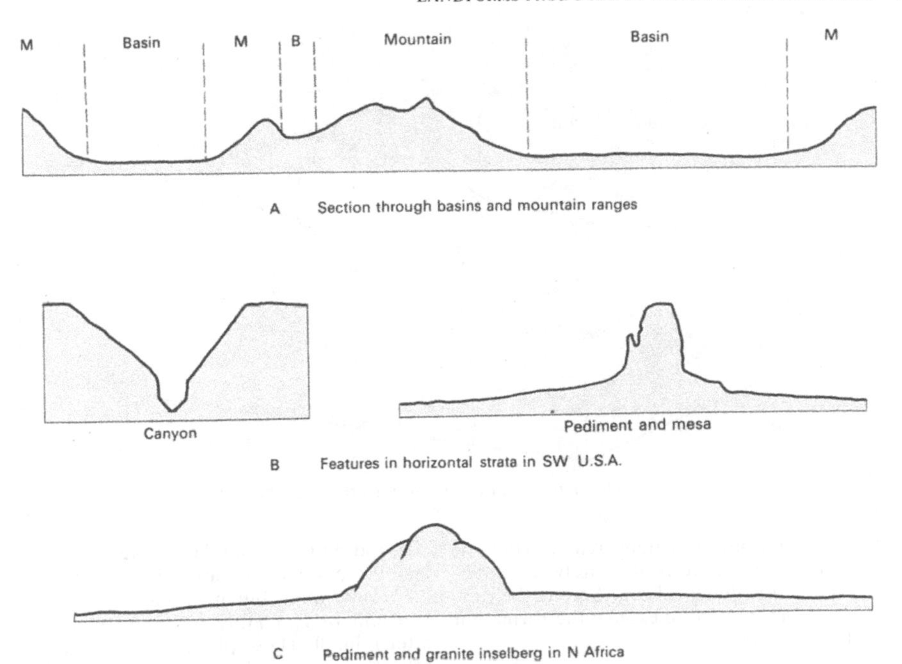

Figure 49 Sections across desert landscapes from North America and Africa.

rounded by flat surfaces (Fig. 49B). In both cases, the mountain areas tend to be the primary location of erosion, and the basin centres are largely affected by deposition. Denudation processes in deserts tend towards the production of a flat surface by mountain erosion and basin infilling.

In the mountain areas, the landscape is dominated by bare rock and debris slopes. The rock outcrops are often eroded into elaborately shaped pillars and mounds by combined weathering and surface runoff. Permanent and ephemeral rivers usually erode very steep-sided gullies or gorges, the steep slopes resulting from the failure of hillslope denudation to keep pace with channel erosion (cf. limestone gorges on page 48). In the south-west United States, these features are called **canyons** (Fig. 49B). The classic example is provided by the Grand Canyon on the Colorado River. In this case, a permanent (and therefore permanently eroding) river flows through a desert of minimum hillslope denudation.

Where a canyon issues from a mountain area into a desert basin, an **alluvial fan** is normally deposited (Fig. 50). The fan is shaped like a segment of a cone with its apex at the exit of the gully or canyon. Deposition of river bedload at this point seems to be due to a number of factors. On issuing from a gully, the river can increase in width, which results in a less efficient channel. Alternatively, the sudden decrease in gradient, coupled often with a decrease in discharge due to infiltration, may cause velocity to fall, which will lead to deposition. The fan shape is probably the result of constantly changing river courses below the point of issue from the mountains.

Between the edge of the mountains and the centre of a desert basin there is usually a flat surface. This surface often meets the mountains at a very marked break of slope (Fig. 50). Towards the basin centre, the surface is usually a depositional feature, an alluvial plain called **bajada**. Closer to the mountains, however, the surface may be an erosional feature cut into bedrock. The term **pediment**, which is used in a variety of ways, can be used to describe this bedrock surface. The pediment seems to lie close to the hypothetical dividing

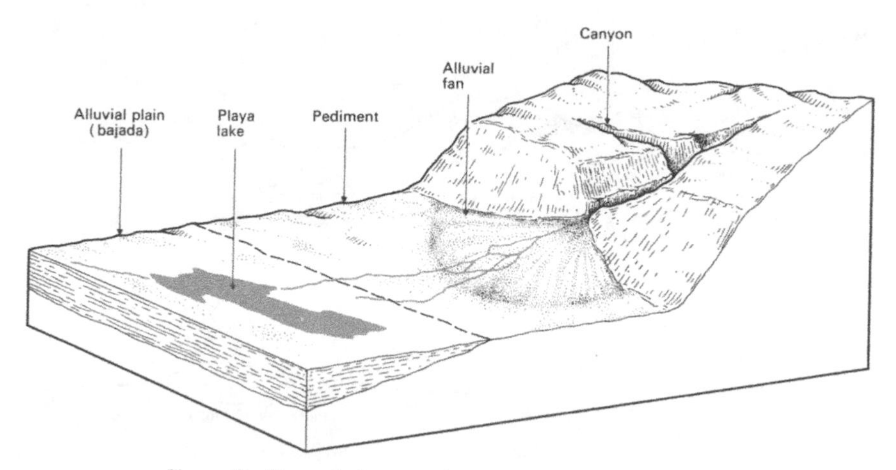

Figure 50 The main features of a water-eroded desert landscape.

line between erosional mountain areas and depositional basin areas. There is absolutely no agreement about how pediments form. Among the various theories put forward to explain the formation of this flat rock surface are:

(a) erosion by sheet floods;
(b) erosion by meandering ephemeral rivers which constantly change position;
(c) weathering which clothes bedrock with a thin, but variable, veneer of debris;
(d) it is the result of the parallel retreat of the steep mountain slope.

On the whole, the final explanation is perhaps the most likely, although it is necessary to find a reason for the sharp break between mountain and pediment slopes at the beginning of the retreat. In areas which are heavily faulted, the original mountain edge may have been a fault scarp.

It would appear that, over time, pediments extend and mountain edges retreat. The basin areas therefore grow. The end product of this process is the relict upstanding block, an **inselberg**, surrounded by pediments. In Africa, many inselbergs are composed of granite, which tends to weather into very rounded shapes called **bornhardts** (Fig. 49C). In the south-west United States, however, horizontally bedded rocks result in flat-topped **mesas** (Fig. 49B).

In the basin of a desert, river sediment accumulates. In the very centre of the basin, the very finest clay and silt is deposited in a lake. Sometimes this lake is a permanent feature if riverflow can meet lake evaporation, but in most desert basins the lake is a temporary feature which does not last long after rainfall. These **playa** lakes (Fig. 50) evaporate to leave a flat plain of fine sediment. If a playa is fed by ground-water as well as by surface streams, the dissolved products of chemical weathering may be precipitated on to the surface. These **evaporite** deposits include a range of salts such as common rock-salt (sodium chloride), various nitrates, sulphates and carbonates. These different salts occur largely in separate layers which may be the result of either precipitation from lake water when saturation of that salt is reached or differing mobility of the various salts upwards through permeable sediments. Famous examples of saline playas include the Great Salt Lake of Utah, and other evaporite deposits, such as the nitrate beds of the Atacama, are economically very important.

E. Transportation, erosion and deposition by wind

The absence of vegetation and the usually dry state of the surface in deserts make wind a far more effective agent of denudation than it can be in humid climates. At the same time, it is easy to overestimate the importance of wind in deserts relative to the effect of water. It should be remembered that whilst wind operates to some extent over all desert areas, 'sand seas' or **erg** actually

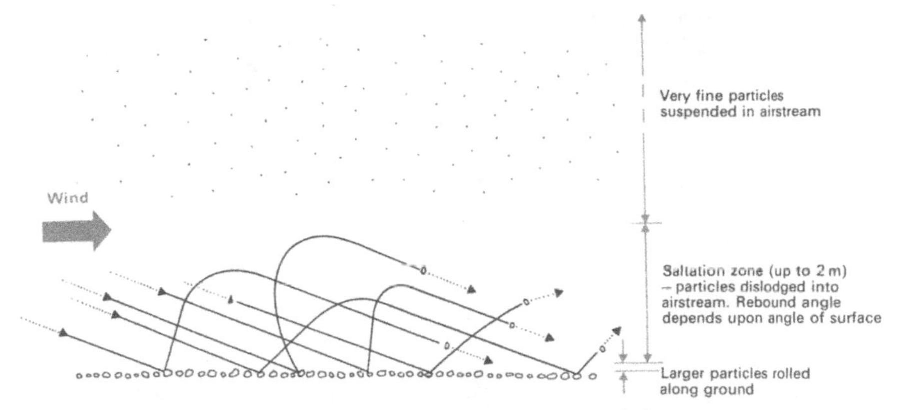

Figure 51 The transport of particles by the wind.

extend over something less than one-third of the arid belt.

Wind transports weathered particles much as does a river. The amount of material carried, for example, increases with velocity. On the other hand, air being less dense than water, wind can generally carry only smaller particles. Only the very smallest silt-sized particles are carried suspended in the airstream to form the dust storms of desert areas. Particles of sand size and above either are moved along the ground surface by creep or are moved by saltation. Surface creep should be compared with bedload movement in a river. The actual process includes a combination of rolling by direct tractive force of the airstream and dislodgement by smaller particles falling to the ground. Saltation is the process by which smaller particles are bounced up into the airstream by the dislodging effect of other material moving at the ground surface and are then carried some way in the faster wind above the ground before sinking back to the surface (Fig. 51). The impact of landing particles is important both in maintaining a supply of particles into the airstream for saltation and in assisting the creeping of bed particles. On average, saltation moves about three times as much material as bed creep.

The direct result of the capacity of the wind to transport material is that erosion by wind abrasion can take place wherever bare rock is exposed: wind is not confined to a channel as is a river. Similarly, while all saltation particles are normally carried within 2 m of the ground, wind is not gravity-controlled in the same way as water or mass movement processes and may therefore carry out erosion effectively on all parts of a hillslope or rock outcrop.

On a very small scale, wind erosion may cause the abrasion of individual pebbles which show highly polished flat (faceted) surfaces. Pebbles of this sort are called **ventifacts**. Individual rock outcrops may be 'sand blasted' by the wind, the surface being rounded and polished. Rock outcrops which are eventually shaped to lie parallel to wind direction by erosion are called **yardangs**. Wherever an outcrop consists of varying layers of rock, the weaker bands are eroded and, because the wind removes the debris in a horizontal direction rather than causing it to pile up at the base of the outcrop, various forms of undercutting are possible. On a larger scale, some areas of desert, such as the Tibesti Highlands in the Sahara, have enormous parallel grooves eroded by the wind. Each groove may be up to a kilometre in width and may extend for many kilometres along the direction of the prevailing wind. Finally, if erosion by the wind is concentrated in one surface area, a **deflation hollow** may result. Deflation hollows may be of almost any size (the Qattara Depression in Egypt, for example, although not entirely of this origin, is 300 km long) and may extend below sea-level.

The centres of desert basins tend to become areas of wind deposition as well as alluvial deposition. Wind does not simply deposit particles in a thick sheet, but also produces many different shapes in the sand which are constantly moving.

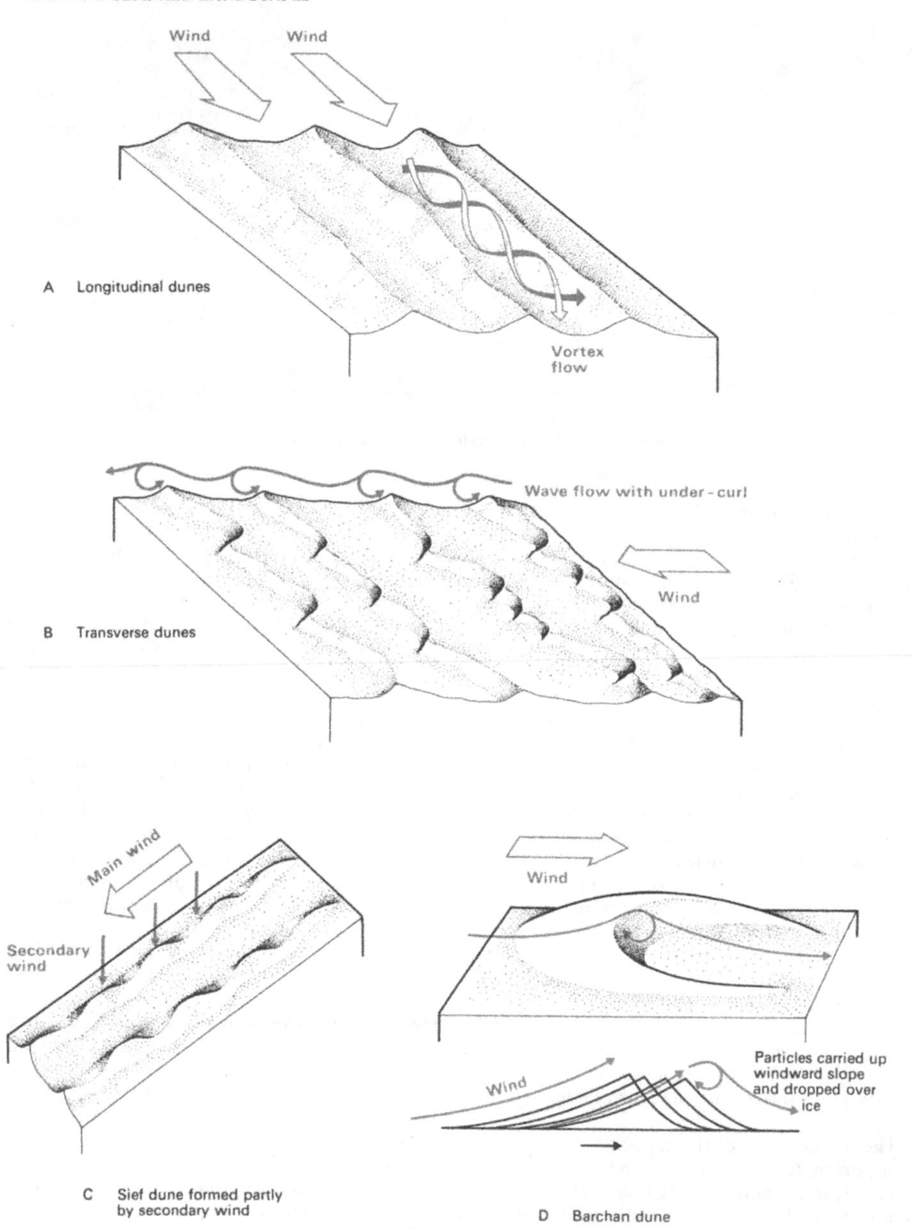

Figure 52 Desert sand dunes.

Because these features are really only 'temporary', the term 'bedform' rather than 'landform' is used to describe them: the 'bed' being the wind/ground contact zone. The names applied to the various bedforms of desert sands are very confusing, but they can initially be classified by size. Since bedforms are essentially 'waves' in the sand, it is possible to describe them by *wavelength* (that is the distance between two sand crests) and *amplitude* (that is the height difference between crest and trough in the sand) (see Chapter 5 on coastal waves). The following groups are defined on this basis:

	Wavelength	*Amplitude*
1 Ripples:	Up to 2 m	Up to 5 cm
2 Dunes:	3–600 m	0·1–100 m
3 Draa:	300–5000 m	20–450 m

All three sizes of bedform can occur within one area: thus, ripples can be found on dunes and dunes can be found on draa.

These three size categories seem to be related to the turbulence eddies within the airstream operating on different scales. Both draa and dunes, and to some extent ripples, may take on a variety of shapes (Fig. 52). The exact shape seems dependent upon such things as wind direction and flow pattern. Both **longitudinal** and **transverse** shapes seem to occur where wind is from one direction but, whereas transverse shapes seem related to a wave form of eddying, longitudinal shapes are developed where the wind forms a series of horizontal and parallel vortices. Where wind comes from more than one direction other forms may result. The **sief** is a longitudinal and sinuous form with a very sharp ridge. The longitudinal element lies parallel to the prevailing wind direction and a secondary wind creates shaped crest forms along the central ridge (Fig. 52C). Where two wind directions produce two alignments of draa, the meeting points may turn into enormous sand mountains called **rhourds**. The same effect can also be produced by one wind direction creating both transverse and longitudinal elements.

Sand is moving in all these bedforms but it is easiest to observe in the **barchan**. Barchans are crescent-shaped dunes or draa which are found only where limited amounts of moving sand cross a hard rock surface. The horns of the crescent point in the direction of movement. Sand is moved up the gentle backslope by the wind and falls over the steep lee of the dune (Fig. 52D).

Finer silt particles which are carried in suspension by the wind may be taken up to hundreds of metres and may not then fall out for several thousand kilometres. When this material is deposited, it forms an enveloping blanket of **loess** which covers an entire landscape. Loess is most commonly found in the semi-arid areas on the leeward side of deserts. Northern China, for example, has extensive loess deposits which originated in the Gobi Desert. Even in Britain, occasional fall-out of red dust from the Sahara can occur under the right meteorological conditions.

Chapter 4

Glacial and Periglacial Landscapes

In glacial areas of the world, much of the water in the landscape is frozen throughout the year. These areas (Fig. 2) include the mountain ranges of otherwise temperate zones and areas within the Arctic and Antarctic circles. Generally speaking, the normal form of ice in temperate zone mountain areas is the **valley glacier**—a river of ice flowing from the highland zones of ice accumulation to the warmer reaches of the surrounding lowlands. In the Polar regions, however, ice tends to accumulate as **ice-sheets** of great thickness and areal extent which melt as warmer water or lower latitudes are encountered.

Although there are enormous differences of scale between these two major occurrences of ice, many of the denudational processes are similar. Consequently, this chapter will examine first the valley glacier system in some detail before looking briefly at ice-sheets. The end of the chapter also includes a section on **periglacial** landscapes which may be found in the areas of Northern Canada and Siberia adjacent to the ice-sheets but not actually covered by ice throughout the year.

A. Water and ice in a valley glacier

Glaciers normally start as accumulations of ice in pre-existing river valleys. Glacier ice is produced by the compaction of snow, the density of snow being 0·1 g/cm³ and that of glacier ice being 0·9 g/cm³. Once a sufficient thickness of ice has developed, the glacier will begin to move downslope under the influence of gravity. The complete valley glacier system will probably consist of a number of high-level tributaries feeding ice to a main valley glacier (Fig. 53A).

A glacier can move downslope in several ways. In the mountains of temperate zones, the ice at the

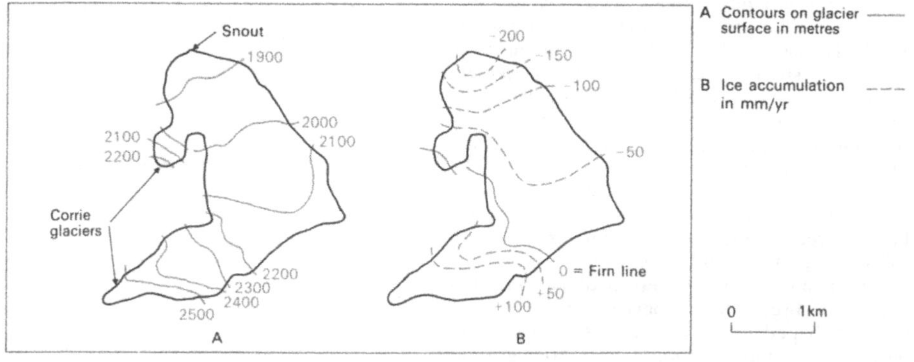

Figure 53 The Place River Glacier, British Columbia: (A) contour map of the ice surface; (B) lines of equal ice accumulation during one year.

bottom of a glacier will melt owing to the pressure exerted by overlying ice, despite the fact that the temperature will be less than 0°C (the 'freezing point' only at normal pressure). Under these conditions the glacier is able to *slide* over the rock surface on this thin layer of water. In very cold climates, it is possible that pressure is insufficient to overcome the low temperatures, and, consequently, the ice remains frozen to the underlying rock. When basal sliding does occur, however, the ice acts partly as a rigid block of material. Further movement can be created within this rigid block by *faulting* and sliding within the ice mass (where it moves over an obstruction, for example). The faulting produces **crevasses** which extend down into the ice from the surface of the glacier. To some extent, ice can also behave in a plastic form, that is it can *deform*, with the result that surface ice velocities are higher than velocities at the base of the glacier. Average glacier velocities are of the order of 10–200 m/yr. Average velocity tends to increase with the gradient of the ice surface and also with the thickness of the ice, since a thicker layer of ice deforms more easily than a thin one. It seems likely that glaciers in 'temperate' regions move mainly by basal sliding, with assistance from faulting and deformation, whereas 'cold' glaciers move more slowly and mainly by deformation and faulting.

Within a particular glacier, the point of maximum velocity and ice discharge is usually at the point of maximum ice thickness, which is found somewhere near the mid-point of the glacier. This situation may be compared with a river, where maximum water discharge occurs at the end of the system. In a glacier system, the **accumulation** of ice occurs at all points on the tributary and main glaciers but is likely to be greatest at the head of the system. At the same time the loss of ice by melting (**ablation**) takes place over the entire system but is inevitably greatest in the lower portion of the system where temperatures are highest. Figure 53B shows lines of equal annual ice accumulation for the Place River Glacier, British Columbia. At the head of the system, annual accumulation is greater than annual ablation. In the lower reaches, the situation is reversed. The **firn line** is that point of equilibrium where accumulation and ablation are equal. Ice moves continuously across the firn line to balance the volumes of ice on the two sides of the line (Fig. 54A). The **snout**, or terminus, of the glacier occurs at the point where ablation is equal to the forward movement of ice. Between firn line and snout, ablation increases, producing **meltwater** from the ice which gradually declines in thickness and discharge away from the firn line.

Seasonal variations have the effect of moving the position of the firn line on the glacier (Fig. 54B). The position of the snout may move seasonally or may remain static while the thickness of ice varies. At the end of a year, the glacier will occupy the same position if total accumulation is equal to total ablation over the glacier area. If this ice balance is not achieved, the snout will advance or retreat absolutely, depending upon whether accumulation or ablation is greater. Over longer periods of time, changes in the climate of an area

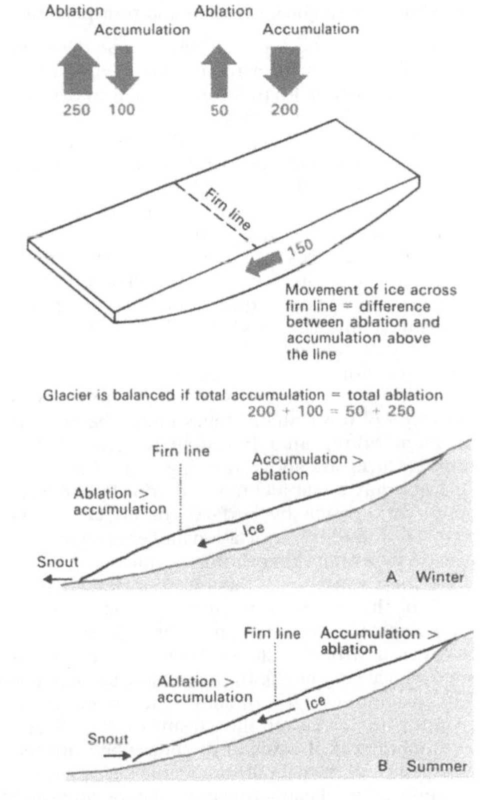

Figure 54 (A) The water balance of a valley glacier. (B) Seasonal movements in the position of the snout of a glacier.

may lead to the large-scale advance or retreat of glaciers.

Under stable conditions, however, the meltwater produced by ablation (especially in summer) is a very important denudation agent in its own right. Ablation produces meltwater which may flow over the ice surface as dispersed flow or in channels, or the water may penetrate the ice through crevasses and other openings to flow through or beneath the ice. This meltwater is capable of both erosion and transportation and may be responsible for landform features beneath and adjacent to the ice itself.

B. Processes of glacial erosion and transportation

In many respects, a glacier carries out the same tasks undertaken by a river. This is particularly true with regard to the dual erosion and transportation role of a glacier.

As a transporting agent, a glacier, like a river, carries the debris of denudation from the adjacent hillslopes. In a valley glacier system, the hillslopes above main and tributary glaciers are likely to be partly bare rock and partly snow-covered. On bare rock surfaces, weathering in the form of freeze–thaw action (see page 21) will occur during periods of temperature fluctuation around freezing point. The products of freeze–thaw will probably be transported to the glacier by rockfall and scree slope processes. In addition to the normal rolling and sliding of scree slopes, rock debris may 'creep' down steep slopes under the encouragement of repeated freeze–thaw cycles. When temperatures are above freezing point, bare rock surfaces may be subject to weathering by chemical action or erosion by surface runoff. On snow-covered slopes, transportation and erosion may be caused by **avalanching** during periods of thaw.

Because ice is solid, this hillslope debris, upon reaching the glacier, comes to rest on the surface of the glacier. In some cases, the entire glacier surface is covered with this debris (**moraine**), giving an appearance very unlike the pure white ice of popular imagination. Frequently the moraine is grouped into ridges on the ice surface (Fig. 55A), forming **lateral moraine** at the edges beneath the hillslopes and **medial moraine** in the centre, which is produced by the meeting of the lateral moraines of two glaciers.

Again, unlike a river, the size or quantity of material which can be carried is unrelated to the velocity of glacier movement. Nevertheless, because the ice surface is broken up with many crevasses and other cavities, some moraine from the surface gradually makes its way into the glacier or even to the base of the ice.

Erosion is carried out by a glacier at the point where ice and rock meet. It is generally agreed that ice is a very effective agent of erosion, but the processes involved are not well understood. The central problem involves what happens at the ice/rock boundary.

For erosion to occur at all, it is clear that the ice must move at the point of contact with rock. It has already been indicated (section A), however, that in very cold climates the glacier base may be frozen to the bedrock, preventing movement. No one is really sure about what happens in these circumstances. In 'temperate' glaciers, where pressure melting occurs at the ice/rock boundary, the glacier moves by basal sliding, and erosion may be caused by two processes:

(1) Abrasion as the result of material entrapped in the solid ice close to the rock. The effect of abrasion may be to polish the rock surface if the trapped material is very fine, or to cause scratches (**striations**) on the rock surface with larger material. It is possible that large trapped boulders may actually cause fresh rock from the bed to be torn out.
(2) **Plucking**. In theory, if ice moves over an obstacle, there will be a slight decrease in pressure on the down-glacier side. Since ice is kept melted at the glacier base by pressure, a slight drop in pressure will cause re-freezing of the basal water. If that water is surrounding fractured rock at the time of re-freezing, the rock will be 'plucked' from its position as the ice moves on.

The total amount of erosion carried out by the glacier will depend upon such things as the pressure and temperature at the base of the ice, since that obviously controls basal sliding, the velocity of ice movement and the amount of abrasive material available at the base of the glacier. The nature of the underlying rock affects the total amount of erosion as it does in a river channel, but the degree of preglacial weathering of the rock is also important. Generally speaking, glacial processes are most effective on already-fractured rock. A particularly important weathering process may be the fracturing caused when erosion unloads the

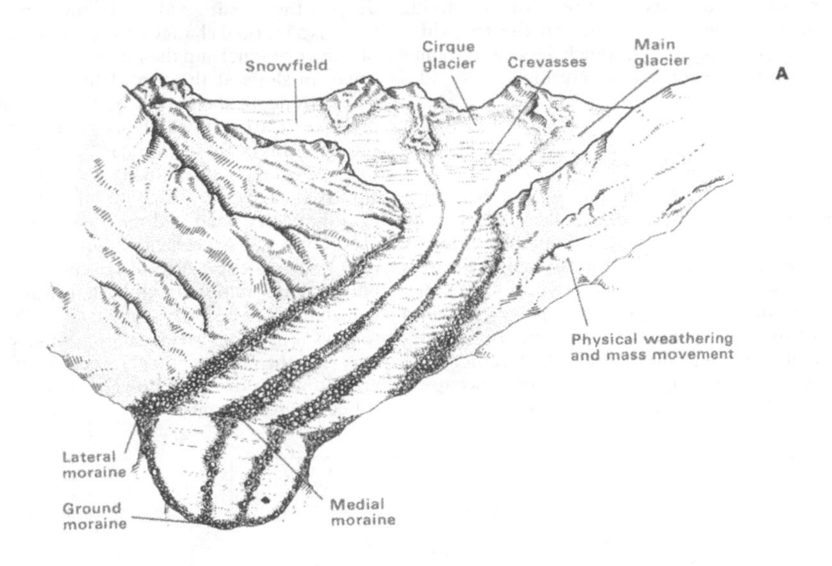

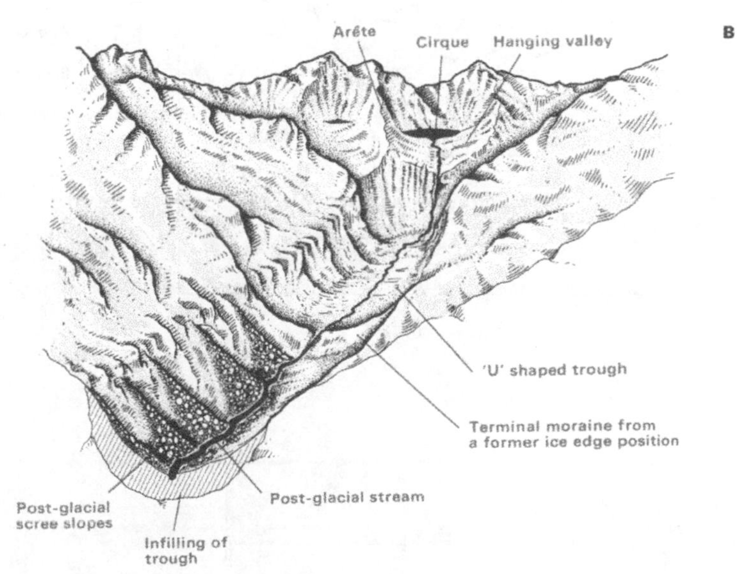

Figure 55 (A) A valley glacier showing ice areas and movement of moraine. (B) The same area after glaciation, showing the main features of glacial trough and tributaries.

pressure on rock at depth. A glacier, for example, can remove enormous quantities of containing rock from the glacial trough, and the ice which replaces the rock will be much less dense, thus allowing pressure release fracturing in the sides and base of the trough.

C. Landforms produced by glacial erosion

The most distinctive feature of glacial erosion is the **glacial trough** (Fig. 55B). Valley glaciers tend to occupy valleys which were originally created by river systems before glaciation. Because ice moves so very slowly in comparison with water, the 'channel' of a glacier is very much larger than a river channel. Whereas the river actually occupies a very small part of a landscape, which is therefore dominated by hillslopes, a glacier may occupy a major part of the pre-existing valley.

By erosion the glacier will widen and eventually deepen the existing valley. In the same way that there is a marked change of slope between the side of a river channel and the hillslope above, there is a break in slope at the top of the glacier ice (Fig. 55B). Beneath the ice itself, the glacier tends to produce a trough which is parabolic in cross-section. This cross-section has often been referred to as 'U-shaped', but some care is needed, since glacial troughs may take on a range of forms, including very deep and narrow 'V' shapes.

Not only does the glacier adapt the cross-section of a river valley; it also alters the plan view. Ice being somewhat less flexible than water, a glacial trough tends to be fairly straight (Fig. 56). Where it follows a previous river valley, the glacial trough may cut through the lower section of the spurs around which the river flowed. Such cut spurs are called **truncated spurs**.

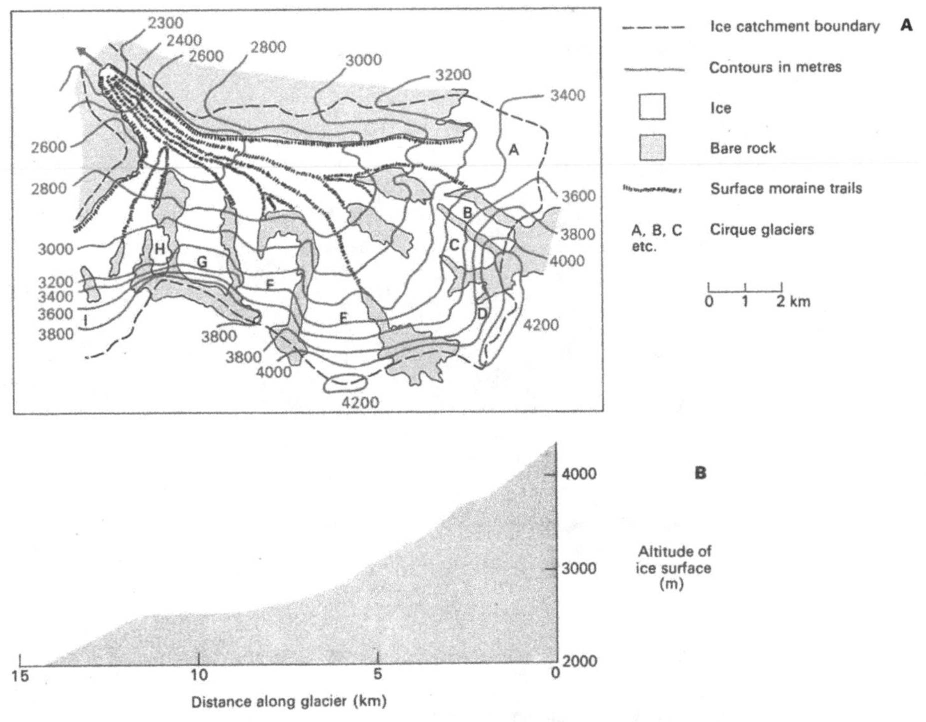

Figure 56 (A) Map of the Gorner Glacier, Switzerland. (B) Long profile of the ice surface of the Gorner Glacier.

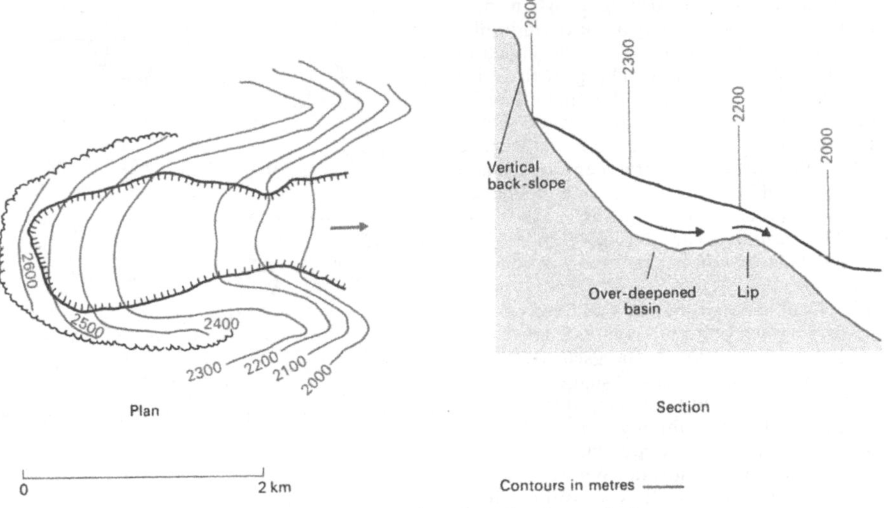

Figure 57 Plan and section of a cirque glacier.

In long profile, a glacial trough is likely to be less regular than a river. Troughs are often broadly concave in long profile, the head of the glacial system being very steep, but the pattern is complicated by frequent steps and concave basins (Fig. 56B). The steps are usually recognised on the glacier surface by ice-falls beneath which active erosion by plucking is taking place. Steps are probably related to sections of more resistant rock. On a smaller scale, blocks of resistant rock in the bed of the trough may be overridden by ice, producing a smooth up-glacier surface and an irregular, plucked down-glacier side. When exposed after glaciation, these blocks are called **roches moutonées**. Where concave basins occur in the long profile, basal sliding may start to act in a rotational manner with faulting occurring throughout the depth of the ice. Rotational sliding of the glacier base may produce enclosed **rock basins** in the bottom of the trough which are deeper than the next section of trough floor (**overdeepening**). Rock basins of this sort often create suitable sites for post-glacial lakes (Chapter 6).

The long profiles of glaciers which are tributary to the main trough often show a marked steep section as the main trough is reached. The glacier of the main trough can generally erode vertically more efficiently than a tributary. The effect is particularly marked after glaciation, when the floor of a tributary trough is seen at some height above the main trough floor (a **hanging valley**, Fig. 55B).

Although valley glaciers normally follow preglacial drainage routes, some departures from that pattern can be seen. If large volumes of ice move into a preglacial valley with a small exit, the level of ice may rise until it overflows the original watershed at a second point. Under these conditions, a **diffluent glacier** is formed. In some cases the diffluent glacier may eventually become more successful and therefore quite alter the preglacial pattern of drainage (see Fig. 96, Chapter 6).

At the head of almost all main and tributary glaciers, ice accumulates in very distinctly shaped hollows (Fig. 56). These **cirques** (called **corries** in Scotland and **cwms** in Wales) are consistently semi-circular in plan and very steep-sided and are often overdeepened beneath the ice (Fig. 57). Cirques may vary enormously in size from tens of metres to several kilometres in width, but the shape remains consistent. In the northern hemisphere, cirques occur mainly on slopes facing north-west, north or north-east, presumably since shade was provided in this aspect for the accumulation of ice.

It is thought that cirques start life as **nivation hollows**. These are high-level hollows in which snow accumulates. In the early stages when snow depth is not great, thawing may occur periodically. During a thaw, water would sink into the cracks in the rock beneath the hollow. The growth of ice during re-freezing would cause the opening up of the cracks so that, at the next thaw, water would seep farther into the rock. At each freezing stage the rocks of the hollow would fragment further and the hollow would gradually become deepened leaving a steep rock wall around the edge of the hollow. The further accumulation of snow (and ice) in the hollow would eventually produce a cirque glacier.

The backs of most cirque glaciers have enormous crevasses which penetrate some distance into the ice below the backface (**bergschrund crevasses**). Some people have suggested that freeze–thaw action might take place at the base of bergschrunds. More commonly accepted, however, is the theory that the cirque glacier behaves rather like a rotational landslide. Where temperatures are sufficiently high to allow pressure melting to occur at the base of the ice, most of the ice movement proceeds by basal sliding, at least until the lip of the cirque is reached, when faulting becomes important as an ice fall is developed. Erosion is affected during rotation by abrasion from backwall material which has reached the bottom of the ice through bergschrunds. Some plucking due to pressure-release freezing may also take place during sliding.

When cirques develop adjacent to one another, the backwalls may intersect to produce an extremely sharp ridge (**arête**). Groups of cirques surrounding a mountain may produce a **pyramidal peak**, such as the Matterhorn in the Swiss Alps, from the intersection of three or more cirque backwalls (Fig. 58).

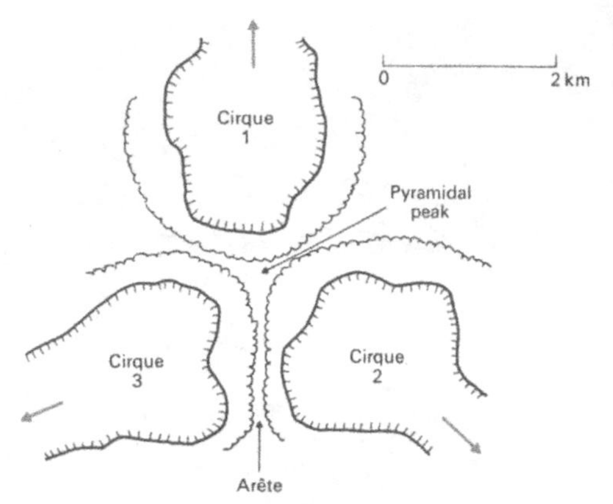

Figure 58 Plan of cirque glaciers around a mountain peak.

D. Landforms produced by glacial deposition

Moraine is carried on the surface, within or beneath a glacier. It is generally deposited as a permanent landform either at the end of an active glacier or along the length of a retreating or stagnating glacier.

The material which makes up moraine is called **till** and is distinctive for several reasons. First, till is **poorly sorted**: that is it consists of a mixture of particles of many sizes deposited together. It is for this reason that some tills are called **boulder clay**. In contrast, most river deposits show only a limited range of particle size at one point, because the size of particle carried depends upon the water velocity. A glacier, however, can carry any size of material irrespective of velocity. Second, till is **unstratified**: that is it does not show the layers typical of water deposits which are produced by a series of separate deposition events such as floods. Finally, the particles in till are very **angular** because they have not been subjected to the rounding processes of moving water which cause abrasion of particles.

The commonest deposit formed along the floor of a glacial trough is **ground moraine**. Typically of the order of tens of metres thick, ground moraine is probably formed beneath stagnating (melting) ice partly from the moraine carried along the base of the ice and partly from moraine on the ice surface which is gently lowered to the ground as the ice melts. Ground moraine does not usually show distinctive features but tends to fill in any irregularities in the underlying rock surface to produce a gently undulating upper surface. **Drumlins** are really part of the same deposit group. Drumlins are areas of 'ground moraine' which have been shaped into streamlined mounds. The typical shape of a drumlin (Fig. 59A) is that of an ellipse, the long axis being parallel to the direction of ice flow, with flattened up-glacier and elongated down-glacier ends. A normal drumlin has no core of solid rock and may consist of till of any particle size range. No really clear theory exists to explain the origin of drumlins, but it does seem important that they

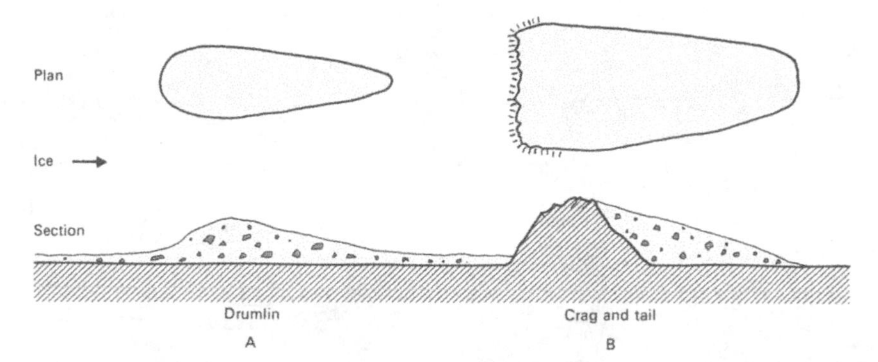

Figure 59 Plan and section of moraine forming (A) a drumlin and (B) a crag and tail.

occur in large clusters (drumlin 'fields') fairly close to the edge of the ice. Drumlins seem to be absent from areas of former thick ice. This pattern has led to the suggestion that drumlins are formed from the deformation of ground moraine by ice flow movements in areas near the edge of the glacier where ice pressure is decreased. The deformed moraine may be initially fairly shapeless. Once in place, however, the ice cannot simply obliterate the deformed moraine, which is therefore streamlined into drumlins by the eroding effect of moving ice. If deformation takes place beneath thicker ice, however, the greater pressures may completely destroy the drumlins before they are fully formed.

A superficially similar feature, known as **crag and tail** (Fig. 59B), is not, in fact, related to the drumlin. The crag is simply a resistant rock feature, such as a *roche moutonée*, and the tail is moraine left on the lee of the crag, probably during the later stages of glaciation when plucking is less effective.

Moraine deposited at the edge of a glacier usually takes on a ridge form. The largest ridges occur at the snout of the glacier as **terminal moraines**. In the case of a valley glacier, the terminal moraine may be shaped around the convex end of the ice and may form a solid wall across the valley (Fig. 60). In a glacier where the snout is stationary, ice will still move up to the terminus although ablation is rapidly increasing. That ice carries moraine which falls forward from the melting glacier edge to form terminal moraine. Where a glacier is advancing, terminal moraines may be formed by pushing existing moraine and other material in front of the glacier into a new ridge (**push moraine**). Finally, it seems likely that ground moraine, when saturated with water, may be squeezed by ice pressure from beneath the glacier into a terminal moraine.

In addition to terminal moraines, **lateral moraine** may be formed along the sides of a

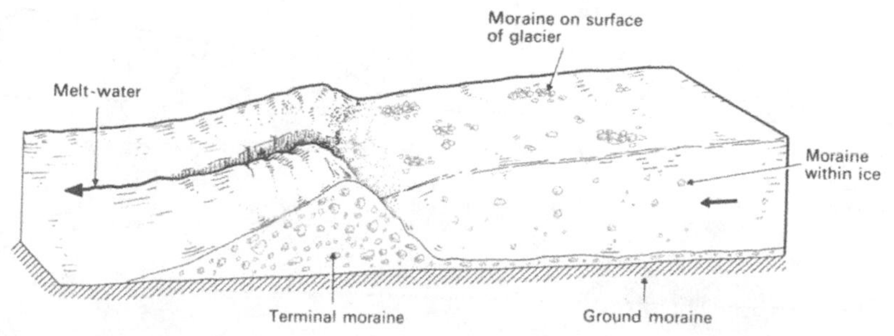

Figure 60 The formation of a terminal moraine at the snout of a glacier.

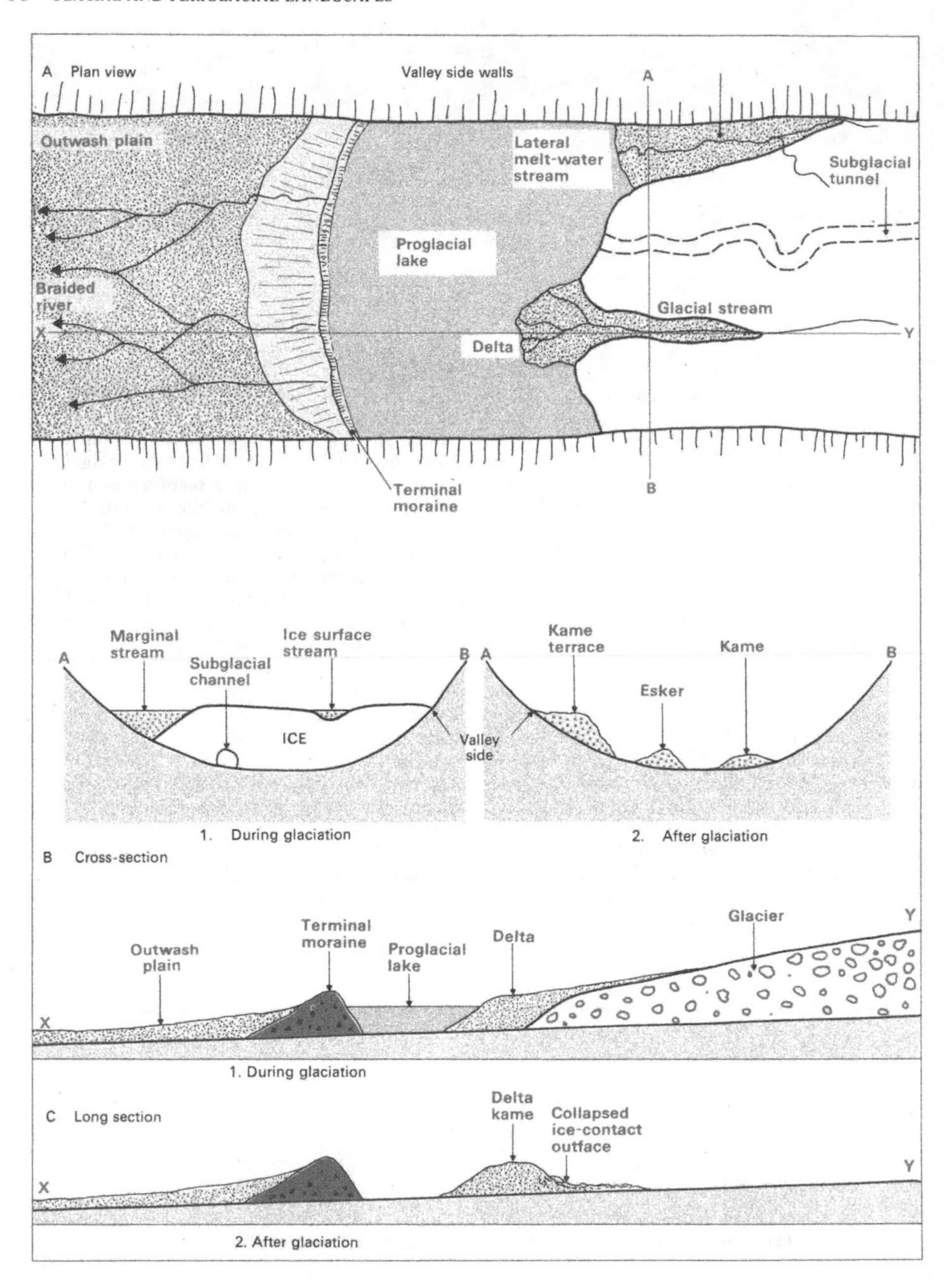

A Plan view
Valley side walls
A
Outwash plain
Lateral melt-water stream
Subglacial tunnel
Proglacial lake
Braided river
Glacial stream
X
Y
Delta
Terminal moraine
B
Marginal stream
Ice surface stream
Subglacial channel
A
B
ICE
Valley side
1. During glaciation
Kame terrace
Kame
Esker
A
B
2. After glaciation
B Cross-section
Terminal moraine
Glacier
Y
Outwash plain
Proglacial lake
Delta
X
1. During glaciation
C Long section
Delta kame
Collapsed ice-contact outface
X
Y
2. After glaciation

stagnating glacier. The moraine in this case will include original lateral moraine from the sides of the ice surface deposited as melting occurs, material from the side slopes above the ice and moraine from the rest of the ice surface. This tends to move to the sides of the glacier during melting, since ice melts more quickly at the sides where it is warmed by reflected solar radiation from rock surfaces.

E. Glacial meltwater

Ablation on the surface of a glacier produces meltwater. The amount of ablation, and therefore quantity of meltwater, increases towards the glacier snout. Enormous variations may occur between the volumes of summer and winter meltwater.

Meltwater may form into channels on the surface of the glacier or at the edge of the glacier. Water entering the ice through crevasses may produce a network of tunnels within or beneath the ice. Many active glaciers have substantial tunnels visible at the snout. Meltwater from the glacier surface may simply flow away from the edge of the ice or may be trapped into lakes next to the ice, where it will accumulate until able to overflow in some direction (see below).

Meltwater inevitably carries a large sediment load reworked from moraine and is therefore capable of erosion when it comes into contact with the ground surface. Distinct channels may be cut by meltwater flowing along the edge of a glacier (marginal channel) or beneath the ice (subglacial channel). The **subglacial channels** are particularly interesting, since they have a number of peculiarities. If a meltwater channel within the ice is following a winding course, it may actually come into contact with the ground surface only periodically. After ice retreat, such a channel will be seen as disconnected sections of channel eroded into the sides of the trough. Furthermore, since water within the ice is often under pressure, subglacial channels may flow uphill for short distances rather as ground water (see page 15) is able to flow uphill.

Beyond the edge of the ice, meltwater flowing down a pre-existing river valley will carry out erosion like an ordinary river except that discharge is seasonal and involves large volumes of water during ice melt, and erosion is therefore more effective. Meltwater rivers like this often show a channel pattern known as **braiding**, where many water courses are constantly splitting and rejoining (Fig. 61A). Where a **proglacial lake** is formed at the edge of melting ice by damming, an **overflow channel** will be cut by escaping meltwater following the lowest route from the area (see page 88 and Fig. 95).

Figure 61 The main features of meltwater deposition showing (A) plan view and (B and C) cross- and long-sections through a glacier snout during and after glaciation.

Although some meltwater channels may be occupied by permanent rivers after glaciation, most of the channels mentioned above will appear in the post-glacial landscape as dry channels which often have a flat floor and steep sides (since slope erosion would be restricted during formation). Most will show a regular downslope course, but subglacial channels may be discontinuous and follow a path which is unrelated to the general gradients of the area.

Although effective as an erosion agent, meltwater primarily affects the landscape through the material it deposits. It may deposit reworked moraine material in two places: in the proglacial area (that is areas close to the ice) and in the ice-contact zone (that is next to the ice itself whether above, beneath or in front of the glacier).

In the proglacial area, where the glacier feeds meltwater into a normal valley, the meltwater streams spread sheets of material across the valley floor to form **outwash plains** (Fig. 61A) of sand and gravel. These plains are often highly irregular surfaces across which the braided streams constantly rework the sediment. Occasionally, blocks of ice from former ice cover may become trapped in the outwash material and later melt to form enclosed water-filled depressions called **kettle holes**. If meltwater streams from the surface of the glacier change to a much lower gradient course in the valley, the outwash material may be concentrated into a **fan**. In all cases, the outwash material will normally be better sorted than morainic material and will show layering in which each layer represents the material washed out during one season's melt. Individual particles in each layer may be aligned with the direction of water flow.

Where meltwater cannot escape from the edge of the ice, a proglacial lake forms in which sediment is deposited. The most distinctive deposit of a lake environment is the thin, horizontal layers called **varves**. Each layer represents the sediment of one summer's meltwater.

In the ice-contact situation are a confusing range of features which are generally called **kames** and

eskers, although there is very little agreement about the exact terms to be used. On the whole, the word **kame** is used to describe a feature resulting from the accumulation of sediment in water next to the ice front. Because the sediment actually accumulates next to the ice, when the ice melts one side of the accumulation collapses. The commonest forms are the **delta-kame** and the **kame-terrace**. The delta-kame is formed in a proglacial lake at the snout of a glacier. The side of the kame farthest from the accumulation of sediment in water next to the ice front. Because the sediment actually the ice-contact surface later forms a collapsed concavity. In post-glacial areas, delta-kames appear as irregular mounds (which may be several kilometres across) the structure of which is only preserved internally. The kame-terrace is formed by accumulation of sediment in lakes and stream channels along the sides of the glacier (Fig. 61B). Following deglaciation, the terrace collapses towards the glacier trough, leaving an irregular ride along the trough sides.

Eskers are long, sinuous ridges of water-lain deposits. They may be of great length and height, although most examples tend to be less than a kilometre long and only ten or so metres high. It is generally agreed that eskers are formed of deposits laid within the tunnels of subglacial meltwater streams. Deposits would naturally accumulate in such tunnels much as bedload accumulates along the channel of a river.

F. Ice-sheets

At the present day, large areas of the Polar regions of the world are entirely ice-covered. Over the land masses of Greenland and Antarctica, in particular, ice accumulates to a depth of several thousand metres. On this scale, ice completely submerges the pre-existing landscape and it is no longer possible to imagine separate 'glaciers' each with its own ice balance. Instead, ice moves in all directions from the highest point of the ice-sheet. An important consequence is that ice-sheets tend, on the whole, to erode an entire land mass. Distinctive troughs are therefore fairly unusual, although examples can be found of cases where part of the sheet ice moved more rapidly along a pre-existing valley to create something like the trough of a valley glacier. Generally, however, unless the ice-sheet was following pre-existing drainage routes, the effect is to plane down the area into an undulating rounded landscape. In lowland areas, ice-sheets may create an almost featureless **lake plateau**, like that in Finland, which is typified by bare rock outcrops and lakes occupying the site of rock basins where overdeepening occurred.

Present-day ice-sheets terminate largely on the margins of the oceans where the moving ice forms icebergs. Under these conditions, little visible evidence remains of the depositional activities of ice-sheets. During the last Ice Age, however, very extensive areas of Northern Europe and North America were covered by ice-sheets (Chapter 6) and evidence of ice-sheet deposition can be clearly seen.

On the whole, the processes and features of ice-sheet deposition are identical with those of valley glaciers but occur on a very much larger scale. Eroded material is carried largely as ground moraine, since very little of the land surface emerges above the top of the ice to provide surface moraine. That ground moraine is left during ice retreat as a huge spread of till covering the rock surface beneath. Drumlins occur on a much wider scale near the margins of ice-sheets, and individual drumlins may be a kilometre in length. Ice-sheets obviously lack any sort of lateral moraine, since there is no 'side' to the ice, but terminal moraines are found in sub-parallel ridges of great height which extend for many kilometres along the former edge of the ice-sheet. The distance over which glacial material is carried is shown by the existence of **erratics** within the till—that is fragments of rock which can positively be identified as having come from a particular location some distance away. The tills of East Anglia, for example, include erratics from Norway.

Meltwater features are similarly extensive, with eskers running for many kilometres and various types of kame reaching hundreds of metres in height. Proglacial lakes impounded by ice-sheets may be similarly impressive in size, producing areas of many square kilometres underlain by lake and deltaic deposits. The overflow channels from these lakes can form significant valley features and often leave a permanent mark on the drainage pattern of an area (see Chapter 6).

G. Periglacial regions

Periglacial regions are those areas in high latitudes where average air temperatures are below freezing point but ice does not actually accumulate on the surface owing to summer melting when air temperatures are above freezing for several months. In

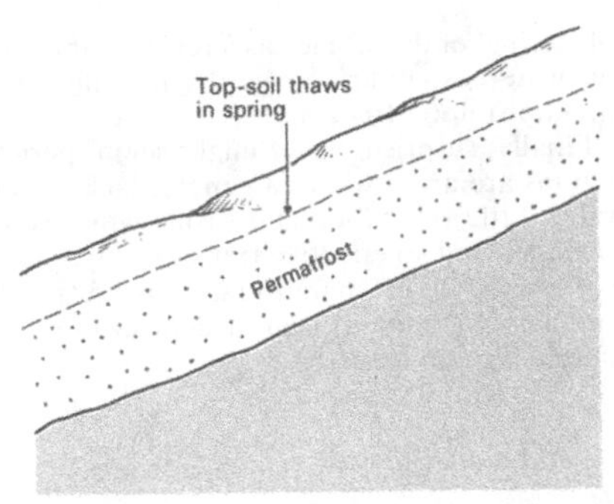

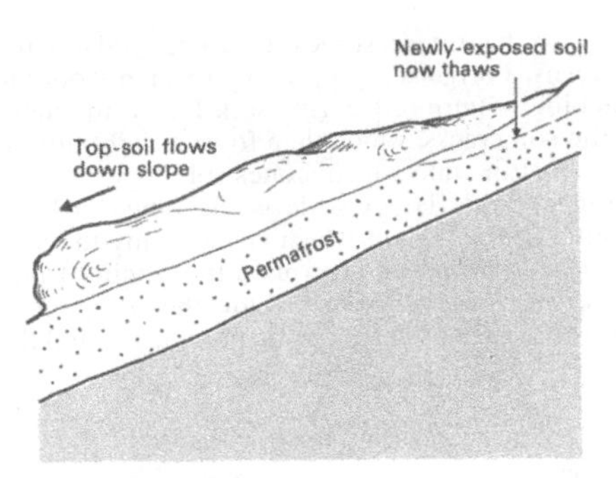

Figure 62 Solifluction on a periglacial hillslope.

periglacial regions, much of the denudation occurs by the processes associated with river systems, although some variations occur. On hillslopes, weathering tends to be dominated by freeze–thaw action, although chemical weathering is by no means entirely excluded. Transport processes on hillslopes include scree and debris flow, but solifluction (see below) is important. Water on hillslopes and in rivers is frozen for much of the year, but may be very effective in erosion terms during the spring and summer months, when most of the annual discharge is released as meltwater. The rivers of periglacial areas are typically braided and carry a large and coarse bedload.

The most characteristic feature of periglacial area, however, is the existence of **permafrost**. This is the ground beneath the surface that is permanently frozen and it reaches a maximum depth of 400 m in parts of Northern Siberia. This permafrost layer is continuous in areas north of 70° but becomes discontinuous southwards. South of about 60°N, permafrost is only sporadic; that is lenses of permanently frozen ground occur in an otherwise unfrozen area. In all permafrost areas, the top layers of soil thaw in summer. The depth of this **active layer** varies, being up to 4 m in the region of 50–60°N in Siberia but decreasing northwards into colder environments.

The effect of permafrost is most noticeable on slopes. When the surface layers thaw in summer, the meltwater cannot percolate vertically into the underlying still-frozen ground and the active layer becomes saturated. Under these conditions, the active layer may move downhill as an accelerated form of soil creep called **solifluction**. Solifluction progressively lowers slopes by exposing new sections of permafrost on the hilltops to thawing (Fig. 62). In the valley bottoms, solifluction deposits may accumulate to some depth. These deposits are usually poorly sorted but often show internal flow structures from their rapid downslope movement.

Permafrost also has some other rather strange effects. The best-known of these is the phenomenon called **patterned ground** (Fig. 63). Patterned ground is formed of stones on the ground surface arranged into either rows (**stone stripes**) or irregular polygon patterns (**stone nets**). Stripes are found on slopes above 4° and seem to be stone nets elongated by downslope movements of slope debris. The cause of patterned ground is not understood completely but seems to be related to two processes. First, stones are heaved up to the surface during freeze–thaw cycles, probably because ice melts beneath a stone more slowly, leaving the stone standing fractionally higher than the surrounding soil during each melting phase.

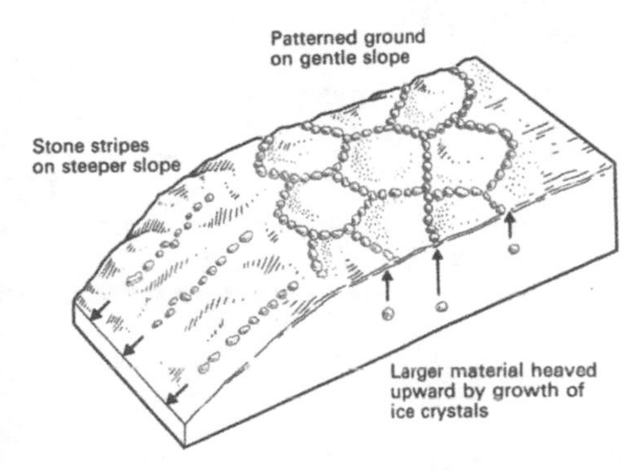

Figure 63 Patterned ground and stone stripes in a periglacial area.

Once on the surface, stones are rearranged into the patterns. Polygons are probably formed from the cracking pattern of freezing soil. Ice accumulates in the soil cracks, which then form a slight hollow after melting into which stones roll.

On a rather larger scale is the **pingo**, a large mound in the core of which is ice. Many theories have been proposed to explain the origin of pingoes, but most of them include the trapping of saturated rock by surrounding permafrost and the 'blistering' of the surface as a result of the remaining water expanding upward (the only possible direction) upon freezing.

Finally, surprising as it might seem, periglacial regions are areas where a general lack of surface water, little vegetation and strong winds combine to make wind an effective transporting agent. The loess of the desert margins (see page 57) is therefore also a common form of deposit in the areas marginal to an ice-sheet.

Chapter 5

Coastal Landscapes

In Chapters 2–4, the processes of denudation in various parts of the world were seen to depend largely upon the prevailing climate. Wherever land meets sea, however, a set of denudational processes operate which do not vary much with the climate of the particular area.

A. Currents, tides and waves

Marine erosion, unlike river erosion, is related not to the amount of water available but to the movements of that water.

From a denudational point of view, large-scale ocean currents are not usually effective agents of erosion or transportation, as their velocity is too low to carry any but the finest particles. Generally speaking, tidal movements created by the gravitational pull of the moon are similarly slow. On the other hand, the tidal range on a particular coast—the vertical difference between low and high tide—will determine the height range over which marine erosion by waves takes place. Tidal range varies seasonally with the relative gravitational alignment of sun and moon and also with the shape of the coastline. Tides tend to be concentrated into funnel-shaped estuaries like the Bristol Channel with the result that the 'height' of high tide increases all the way up the estuary. The velocity of tidal ebb and flow is normally low but will increase with the tidal range and with the horizontal distance which the leading edge of the tide has to cover between high and low water. This latter distance, for a particular tidal range, increases as the coastal profile becomes shallower (Fig. 64). Under some conditions, coastal shape can produce tidal velocities which begin to move material, thus giving rise to **tidal scour**.

Generally, however, the most important movement of water on the coast is the wave. Waves are created by the friction of wind blowing over the water surface. The shape of a wave is described in terms of the **wave length** between troughs and the **amplitude** between trough and wave crest (Fig. 65). As the ratio of wave length to amplitude decreases, the wave becomes steeper (Fig. 65B). Waves become steeper as the wind speed increases, the duration of persistent wind increases or the distance of open water over which the wind has operated (the **fetch**) increases. On the southern coast of England, for example, waves from due

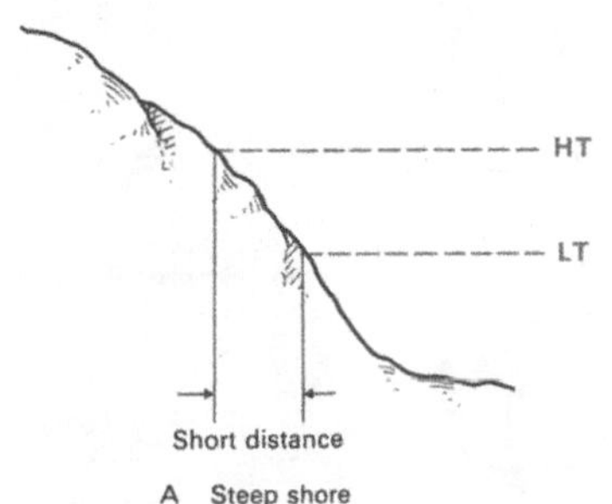

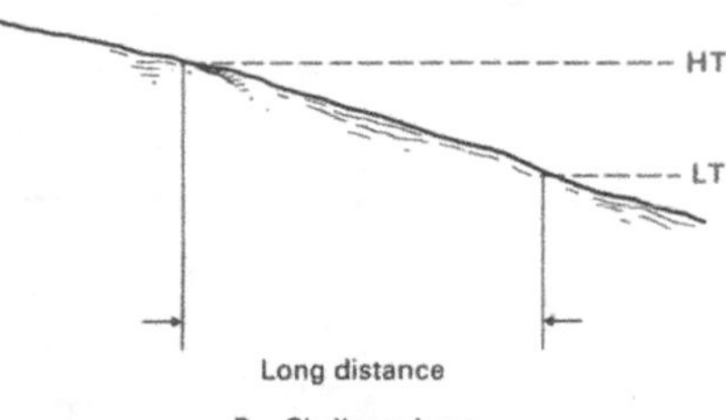

Figure 64 The relationship between shore gradient and speed of tide in two situations with the same tidal range.

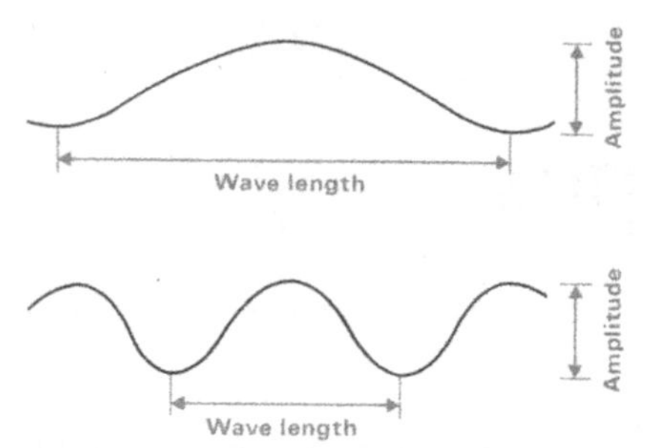

Figure 65 Two waves with similar amplitude but contrasting wave length.

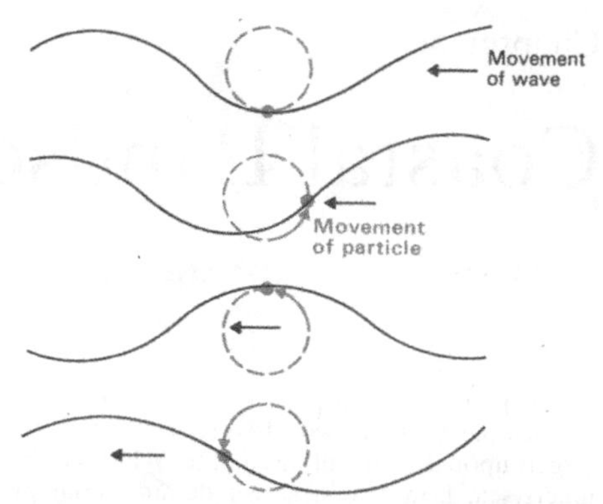

Figure 67 The circular movement of a particle of water with the passage of a wave.

south (Fig. 66) have a fetch of only 200 km compared with the fetch of several thousand kilometres for waves from the south-west. The wave fetch for a particular point on the coast changes as the wind direction varies, but, in addition, different points on the coast will be susceptible to different fetch patterns. The east coast of England, for example, can never be affected by waves which have the fetch of south-west waves reaching Cornwall or Devon.

Within a wave, there is normally no wholesale forward movement of water. Instead, water particles rotate as a wave passes (Fig. 67), producing a slight backwards/forwards motion. As a wave reaches shallower water, however, the bottom of the rotation begins to be affected by drag on the sea-floor. At the point where the average depth of water is half the measure of the wave length, the wave begins to break and the water contained within the wave form moves forward on to the shore.

B. Marine erosion

Above high water mark, the coastline is subject to the processes of denudation on hillslopes discussed in the preceding chapters. Below high water mark, marine erosion is dominant.

Marine erosion, like river erosion, includes a number of processes. The chemical action of sea-water (corrosion) will take place wherever rock

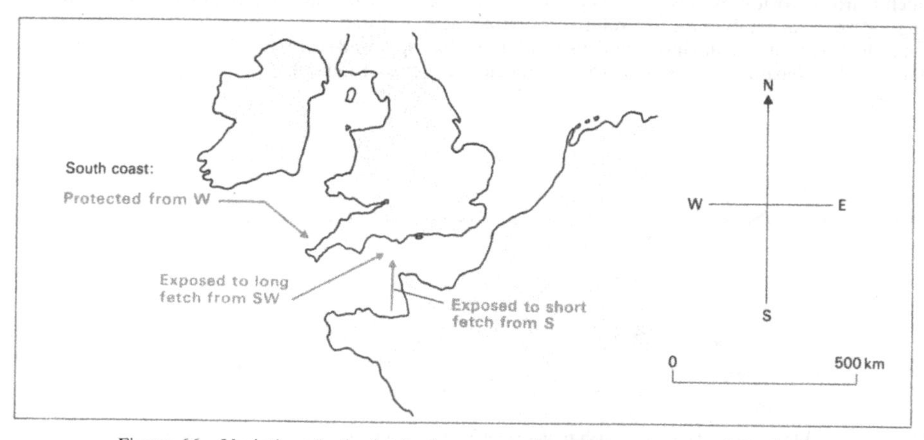

Figure 66 Variations in the fetch of waves reaching the south coast of England.

and water are in contact but will probably be most important in the zone beneath the active wave zone, where physical erosion is reduced. In the intertidal zone, physical weathering by wetting and drying or salt crystal growth may be locally important. The physical action of waves, however, is the dominant erosion process on the coastline.

Wave erosion is largely of two kinds. Simple **hydraulic pressure** can include all the pushing, pulling and compressing of a breaking wave. Notable, perhaps, is the tendency for waves to enlarge cracks in rock by compressing air and water caught within as the wave breaks. Equally, however, the tractive effect of several tonnes of moving water is considerable. The second process is the **abrasive** (corrasive) effect against the rock of particles carried by the waves. In comparison with bedload movement in rivers it is worth remembering not only that waves can carry very large particles, but also that the particles are actually thrown at some velocity against the rock rather than being simply rolled along.

Although some erosion is always taking place on coastlines, much of the distinctiveness of coastal landforms depends upon whether the waves then remove or deposit that eroded material. For any one point on the coast, removal or deposition may occur depending upon the particular form of the waves received at any one time. Generally speaking, waves with a large wave length:amplitude ratio break on to the shore with a forwards movement (the **swash**, Fig. 68) which carries material up the beach. The long interval between these waves allows the return movement (**backwash**) to be completed before the next wave breaks. The overall effect of such waves is therefore **constructive**, since they tend to deposit material on the shore rather than remove it. On the other hand, steep waves with a small wave length:amplitude ratio tend to plunge on to the shore, creating a strong **undertow** in the backwash which pulls material from the shore. Furthermore, since these waves arrive at the shore in rapid succession, the swash of one wave may be rendered ineffective as a construction agent by being forced to flow over the backwash of the preceding wave. Under these conditions, a continuous strong undertow may be established and the waves are then effectively **destructive**.

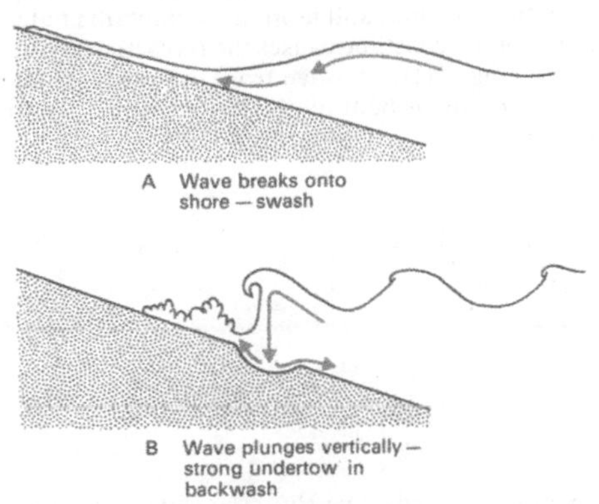

Figure 68 The contrast between (A) 'constructive' and (B) 'destructive' waves on a beach.

All points on the coastline receive waves of both types at some time or another. It is true to say, however, that *on average* some sections of coastline may receive more of one type of wave than the other. **Erosional coastlines** are therefore more susceptible to destructive wave action, owing to exposure to winds of long fetch, perhaps, than **constructional coastlines**. Taking the British Isles as a whole, the western coastline tends to be dominated by erosional features and the eastern by constructional features, although there are many local variations resulting from the detailed pattern of exposure and shelter in the configuration of the coast.

C. Coastlines where erosion is dominant

Wave action on an erosional coastline causes the undercutting of coastal slopes. Where the rocks are fairly resistant, or highly permeable, normal hillslope denudation tends to be slow and undercutting causes rockfall to produce a vertical **cliff** (Fig. 69A). The point of wave attack is sometimes preserved in resistant strata as a **wave-cut notch**, since collapse from undercutting is not always immediate. Where the rock is less resistant (Fig. 69B), rapid undercutting may encourage the development of mass movement processes such as mudflows and landsliding (Fig. 26) which tend to obscure the point of undercutting by supplying fresh material to the shore.

Wave erosion is effective only within the wave zone (through direct attack) or above the wave zone (through collapse of overlying rock). Erosion by these processes does not occur below the level of the wave base. Consequently, the progressive retreat of a cliff by collapse above a wave-cut notch

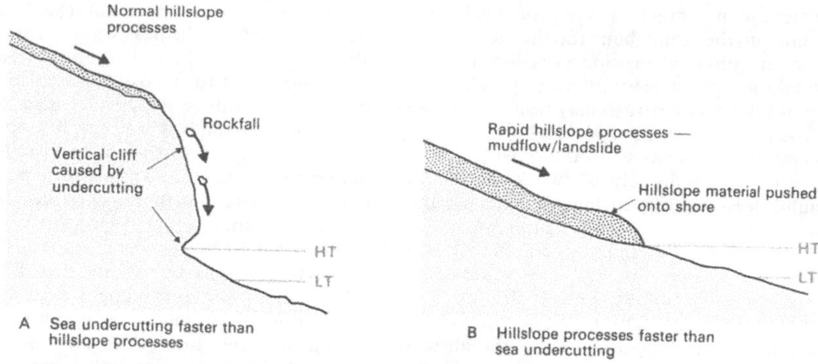

Figure 69 Cross-sections of shorelines where undercutting by waves is (A) more effective and (B) less effective than erosion of the cliff by hillslope processes.

leaves behind a fairly flat surface called a **wave-cut platform** produced by the successive positions of the base of the wave-cut notch (Fig. 70). As cliff retreat continues, the wave-cut platform creates an increasingly wide belt of shallow water. This platform will tend to cause drag on incoming waves at various times, especially at low tide. Eventually, much of the potential energy of waves will be consumed before the cliff base is reached, undercutting and further retreat will decrease and the cliff angle will decline as hillslope processes come to match rates of undercutting. This system is therefore self-adjusting—in other words, there is probably an inbuilt limit to the extent of a wave-cut platform in a particular site.

Unlike a river, which is eroding the landscape towards the level of the sea, the sea itself has no fixed point to which it is working. Like the wind in a desert, the sea will tend to pick out and emphasise all the minor geological variations on the coast. In particular, **bays** are eroded along lines or beds of weakness, leaving the more resistant areas as **headlands**. The phrase **differential erosion** is sometimes used to describe this selectivity in the erosion of weaker points. It should be noted, however, that differential erosion of this type will not continue indefinitely to exaggerate the coastal indentation, since the bay/headland situation is another self-adjusting system. To be specific, the greater the indentation made by a bay, the slower becomes the rate of erosion in the bay, since wave action becomes increasingly less effective. There are two reasons for this decrease in wave efficiency. First, a bay receives waves from a more restricted fetch range than does a headland (Fig. 71A). The deeper the embayment, the more restricted this range becomes. Second, and more important, the indentation of the coastline causes the **refraction** of wave fronts (Fig. 71B). A wave front originally parallel to a shoreline is bent by the indentations, causing

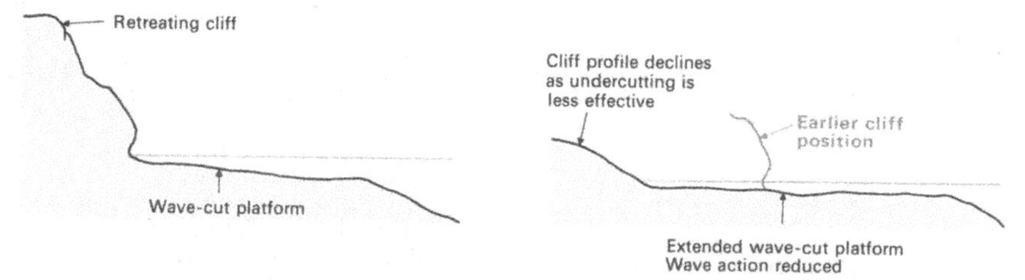

Figure 70 The formation of a wave-cut platform by a retreating cliff. The cliff profile declines as the platform increases in size and makes wave undercutting less effective.

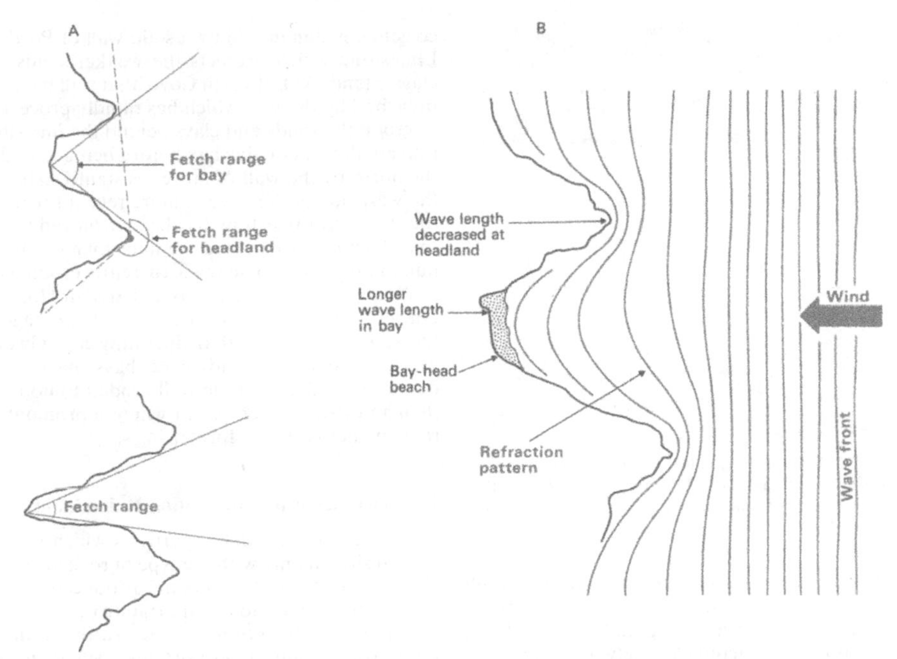

Figure 71 (A) The range of fetch for a particular point varying with the geometry of the coastline. (B) The refraction of wave fronts by a headland to create waves of varying shape.

wave length to diminish at the headlands and to increase within the bays. This pattern obviously encourages headland erosion but causes the waves in the bay to lose erosional power.

The British coastline has many examples of differential erosion. The simplest cases occur in a single rock type where the weaknesses are single faults or large joints in the rock. Around Start Point in Devon, for example, many lines of weakness are developed into small bays (Fig. 72). On headlands such as Ballard Down, Dorset, faults in the Chalk are developed into **sea-caves**. The meeting of two sea-caves developing along a single fault from opposite sides of a headland produces an **arch** which will eventually collapse to leave an isolated **stack** in the sea (Fig. 73).

On a larger scale (Fig. 74), bays may be eroded in weaker beds of rock. In the case of Swanage Bay on the east side of the Isle of Purbeck, the sequence of rock beds lies transverse to the coastline. The sands and clays of Cretaceous age have been eroded more readily than the ridge of Jurassic limestones south of Swanage Bay and the Chalk to the north. Limestone and Chalk now form headlands. Only a few kilometres farther west, the

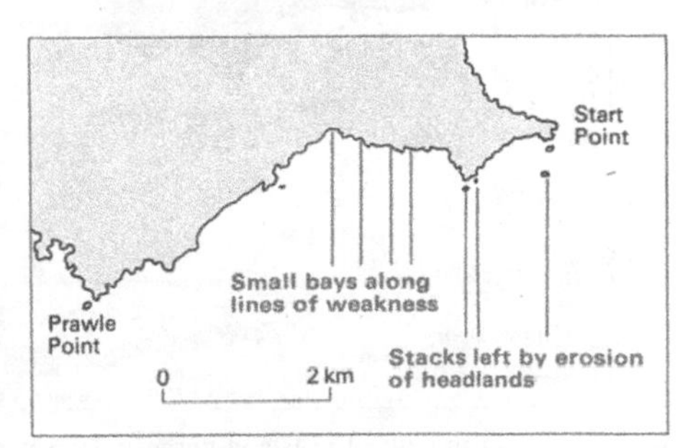

Figure 72 Map of the Devon coast near Start Point, where small bays are eroded along minor faults in the otherwise resistant rock.

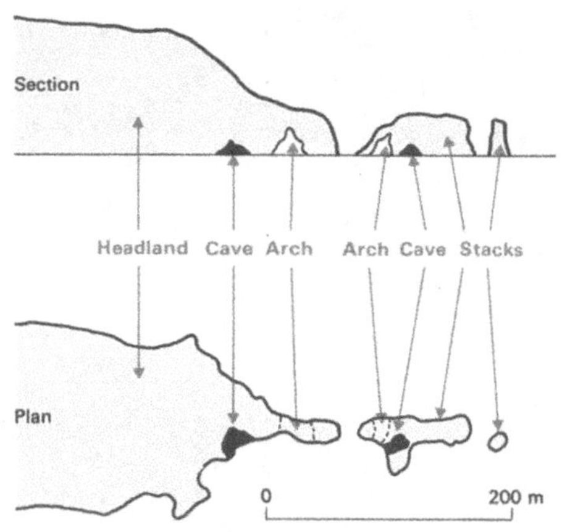

Figure 73 Plan and section through Old Harry rocks, north of Swanage, Dorset, showing the development of caves, arches and stacks by the erosion of faults in the Chalk headland.

coastline runs east–west (Fig. 74), with the result that these same rock outcrops, now somewhat contracted in width, run almost parallel to the shore. In the immediate vicinity of Lulworth Cove, the coastline is dominated by a solid wall of Portland Limestone which protects the weaker sands and clays inland. At Lulworth Cove, that wall has been breached by the sea, which has rapidly proceeded to erode the sands and clays behind the limestone into an almost circular bay before being halted to the north by the wall of more resistant Chalk. On the west side of the cove, a more recent breach of the Portland Limestone has been achieved by the sea which has penetrated the limestone through a number of arches. The sea is currently eroding the sands behind into a small bay called Stair Hole. At Durdle Door, farther west again, a later stage of the same process as that operating at Lulworth may be seen: two adjacent bays have been developed, the limestone wall eroded to no more than an offshore reef except where a promontory remains between the former bays.

D. Coastlines where deposition is dominant

Marine erosion produces particles which initially vary in size, mainly with the type of rock available. Where very large boulders are produced by erosion on headlands, no transportation of the eroded material is likely until it has been further reduced by marine action. Transportation will then take

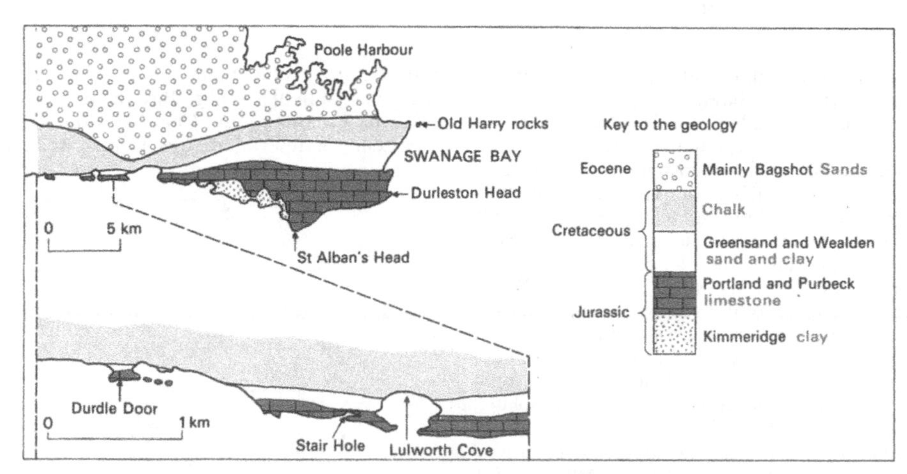

Figure 74 The Isle of Purbeck, Dorset, showing coastlines formed by differential erosion of a rock sequence. Rock strata lie transverse to the shore on the east coast and parallel to the shore on the south coast. The inset shows an enlarged view of the Lulworth Cove area.

place at a rate dependent upon the size of the material and the energy of the transporting waves.

The commonest area of marine deposition is the **bay-head beach** (Fig. 71), where material transported down the wave front from the headlands joins material eroded from the bay-head. Material at the bay-head is likely to be trapped at that point for good although it will, of course, gradually be reduced in size. Taking the beach as a whole, particle size will depend upon such factors as local rock type (clay, for example, obviously breaks into smaller particles than sandstone), geology of adjacent headlands, distance to adjacent headlands and length of time elapsed since original erosion. Taking beaches in general, sand is overwhelmingly the commonest material, followed by shingle and cobbles. Silt- and clay-size particles are rare on a beach, since they are too susceptible to removal by the constant action of waves. Instead, fine particles accumulate in more sheltered locations (see below).

Most beaches show a certain amount of spatial variation. Beach profiles, for example, often include a number of ridges of increasing height, ending at the **berm** or crest of the beach. Each ridge can be probably related to a storm height and they are therefore called **storm ridges**. The berm itself may be well above high tide level and may be produced by material thrown upwards in spray during storms. The storm ridges tend to include the largest particles which have remained stranded since the storm which emplaced them.

Movement of eroded material may also take place *along* the coast. When wave fronts hit a shoreline at an angle (Fig. 75), the forward movement of water and material is diagonal to the beach. The backwash, however, will simply follow the line of maximum slope which is back to the water's edge by the most direct route. The result is that a lateral movement of material called **long-shore drift** is initiated. The rate and direction of long-shore drift will vary around the coastline, depending upon the direction and effectiveness of the winds; that is the combination of direction, frequency and fetch. On the British coast, the most effective winds in terms of both wave frequency and fetch are from the south-west, with the result that long-shore drift operates from west to east along the south coast and from south to north along the west coast. The east coast being in the lee of south-west winds, it is the north-east winds which tend to promote long-shore drift, in this case a weaker drift from north to south.

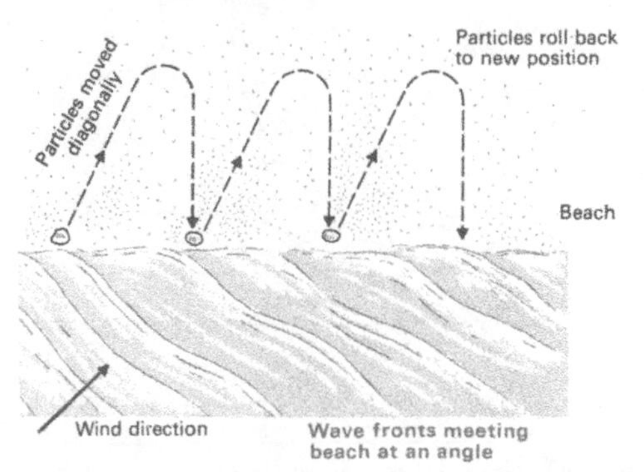

Figure 75 The lateral movement of beach material by long-shore drift on a coast where waves meet the shoreline at an angle.

One effect of long-shore drift is to supply material to beaches far along the coast from a headland which is being eroded. Normally, these **lateral beaches** on a straight coastline are also supplied by erosion from the back of the beach under storm conditions. In many cases, seaside towns such as Brighton have erected wooden groynes across their beaches in order to collect more sand being transported along the coast by long-shore drift (Fig. 76). The cliffs of the seaside towns are normally also protected against erosion by a seawall and promenade, and some beaches on the south coast are deteriorating, since both sources of material supply have now been stopped.

An extension of the long-shore drift process may produce a **spit**. Spits are usually a continuation of a lateral beach across some large embayment in the coast such as an estuary. Hurst Castle Spit in the Solent (Fig. 77) has probably been formed from material carried around Christchurch Bay by

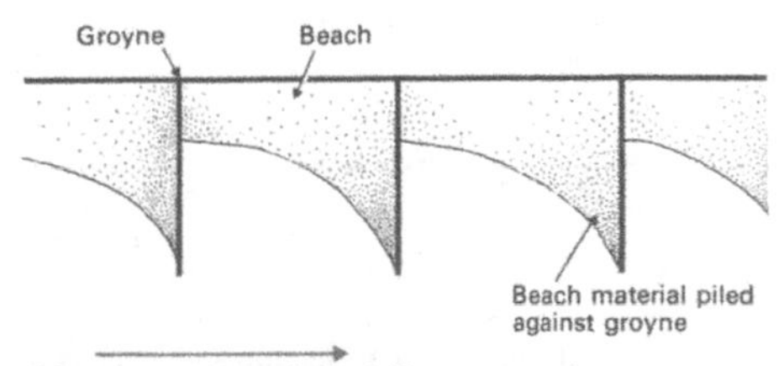

Figure 76 The accumulation of beach material behind man-made groynes as a result of long-shore drift.

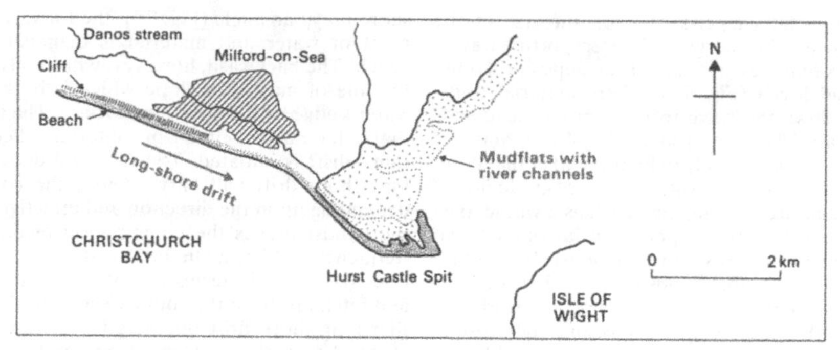

Figure 77 Hurst Castle Spit, Hampshire, created by the long-shore drifting of beach material from Christchurch Bay.

long-shore drift. The sudden change in direction of the coast to the north-east beyond Milford-on-Sea would result in gradual deposition on the bend which built up and became self-perpetuating by extending the line of long-shore drift. The eastern end of the spit shows a number of shingle ridges extending north-west from the main spit. These **recurved ridges** probably result from phases of spit growth during which the end of the spit was regularly being reworked by storm waves moving down the Solent from the north-east.

Occasionally, examples can be found of cases where a spit has grown completely across an embayment to form a continuous **bar** as at Slapton Sands, Devon (Fig. 78A). Such a situation can arise, however, only when no major river enters the sea at the embayment. More generally, a spit growing across a river's mouth will divert the river (for example, the Alde at Orford Ness; Fig. 78B) but will not close its exit.

A rather different process is used to account for **off-shore bars**, which may, like Chesil Beach in

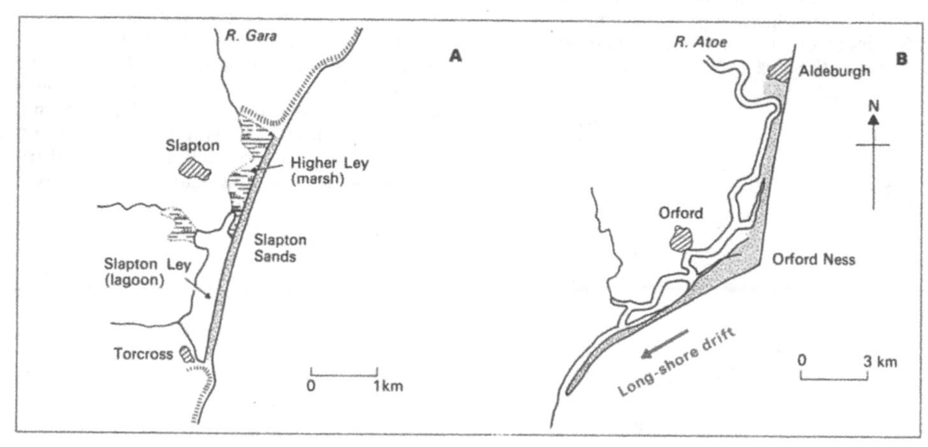

Figure 78 (A) Slapton Sands, Devon, a bar across a number of river valleys which encloses a lagoon (Slapton Ley), now silting up. (B) The diversion of a river mouth by the growth of a bar in the direction of long-shore drift: Orford Ness on the Suffolk coast.

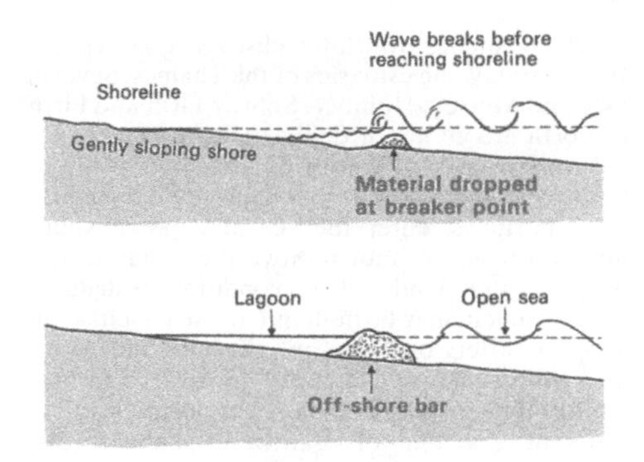

Figure 79 The formation of an off-shore bar on a gently shelving shore where waves break before reaching the shoreline.

Dorset, be attached to the land at one or more points or, like Scolt Head Island in Norfolk, be completely free of the land. The area of water behind such a bar is called a **lagoon** whether or not it is open to the sea. Off-shore bars are thought to originate in the movement of material from beneath the water level on to the shore. An important condition for off-shore bar formation is a gently sloping submarine surface with a cover of eroded material (possibly carried there by off-shore currents). Waves moving across this shallow water covered surface can carry the eroded material forward. Because the water is shallow, waves are liable to break some distance away from the water's edge (Fig. 79) and material is deposited below water level at this point. Over time, the existence of the underwater obstruction encourages further material to accumulate as waves break against the incipient bar. Eventually, the bar emerges above water level and then grows like a normal beach.

Inevitably, in any particular situation both onshore and long-shore movement may take place. In the case of Chesil Beach (Fig. 80), for example, the situation is quite complex. The beach probably started as an off-shore bar created during rising sea-level after the last Ice Age when the water of Lyme Bay was shallow. Prevailing south-west winds created waves which pushed the material lying on the bottom of the bay towards the north-east until it touched land. Once in existence, the bar would grow as sea-level rose. Today, the material on the beach shows a lateral grading which must imply the action of long-shore drift. At the western end (Burton Bradstock), the average pebble size is less than 1 cm. Pebble size increases steadily eastwards until an average of around 7 cm is attained at the eastern (Portland) end. This grading obviously implies a long-shore movement of small material from east to west, which is, of course, counter to the general long-shore drift of southern England. The normal explanation for this phenomenon is that south-west winds produce onshore movement of waves and material but no long-shore sorting, since the beach lies across the prevailing wave front. Instead, sorting is carried out by waves from the south which have a longer fetch, and are therefore more effective, than waves from the west, which is the only remaining direction which could produce long-shore movement. South winds moving on to a beach aligned NW–SE will cause a south-east to north-west long-shore movement. However, this explanation does not answer all the questions. In particular, most of the material on beach is derived from rocks found to the *west* of Abbotsbury. The explanation still does not take account of this problem. It may be, for example, that movement of material from west to east takes place off-shore or that some south-west storms do, in fact, cause eastwards movements of material of all sizes the smaller of which are then reworked back westwards by the weaker south winds.

As indicated earlier, fine material, in the silt and

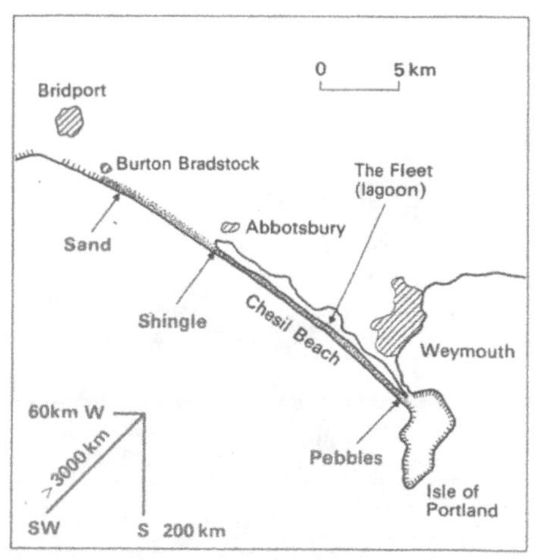

Figure 80 The increase in size of beach material to the south-east along the Chesil Beach, Dorset.

clay range, is not usually deposited in the very active environments described above. Much of the finer material is carried some distance out to sea before being deposited in deep, still water. Some, however, may find its way to very sheltered areas along the coastline. Environments of this type are most frequently found in the lee of a coastal promontory (on the north side of the Gower Peninsula, South Wales, for example) or in the lee of a depositional feature such as Hurst Castle Spit (Fig. 77). These areas of accumulation usually start as **mudflats** exposed only at low tide but, if they are colonised by salt-tolerant vegetation, accumulation will develop **salt marsh** which can be reclaimed eventually. The area immediately east of Milford-on-Sea (Fig. 77), for example, has been created in this way.

E. Deposition by rivers in coastal areas

The flow characteristics of a river are obviously completely altered as it enters the sea, and its suspended load will be correspondingly affected. If a river enters the sea on a straight and active coastline, the river's load may be immediately carried into the marine transportation system. Frequently, however, the point of entry is an estuary produced by 'recent' sea-level rise (see page 84). In an estuary, the conditions are found where all but the very finest particles may be deposited in mudflats similar to those discussed above. In Britain today, the estuaries of the Thames, Severn, Dee and Mersey, Humber, Solway Firth and Firth of Forth are all major depositional areas.

In some parts of the world, very large river systems, carrying a heavy sediment load with fairly large particles, enter the sea at a point where marine erosion cannot remove the material into deeper water. Under these conditions a **delta** of river sediment may be built into the sea. Deltas can adopt a variety of forms but all show some common characteristics. The top of the delta consists of a fairly flat surface (Fig. 81) over which the river flows in a system of branching channels (**distributaries**). The coarsest river load is dropped in these channels as discharge falls by dissemination. High flows spread sand over the rest of the delta platform. Where the channel finally meets the sea, sand is spread on the immediate submarine slope and finer particles are carried out into deeper water. Silt drops out of the water to form a very steep slope around the edge of the delta. Clay-sized material usually stays in suspension until it reaches the deeper water beyond the edge of the delta. Large deltas include the **bird's foot** type of the Mississippi (Fig. 81), where the leading edge of the distributaries protrudes into shallow water, and the more compact **arcuate** forms, like the Nile, which have a regular leading edge.

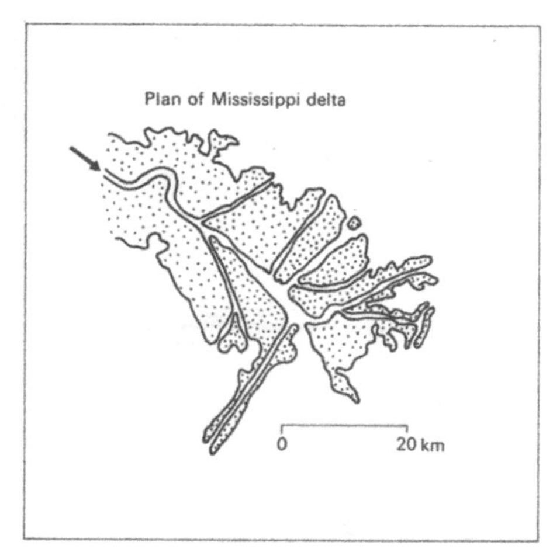

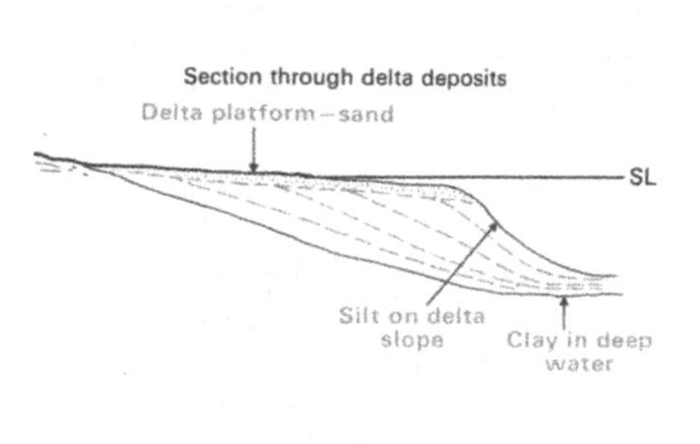

Figure 81 Plan of the Mississippi delta, USA, with cross-section showing deltaic deposits.

Chapter 6

Landscapes of the Past

In Britain today, it is possible to observe the denudational processes associated with rivers, hillslopes and the sea. Many of these processes produce the familiar shape of the landscape. On the other hand, there are features in the landscape which do not seem to be related to the erosional processes which operate in the landscape today. Some of these features are 'left-over' from a past time when the processes were different. There are two main reasons for changes in the erosional processes:

1. In the past, the relative level of land and sea was not the same as it is today.
2. In the past, the climate of Britain was at times hotter and at times colder than it is today.

A. Movements of land and sea

The relationship between land and sea can change either with movements of the land or changes in the level of the sea.

Movements of the land are caused mainly by the mountain building processes referred to in Chapter 1. Deep-seated earth movements over millions of years can cause mountain chains like the Alps or Himalayas to form, but the same forces will also give rise to continuous small changes in almost all parts of the world. The level of the land surface will also rise as a result of erosion. The continents are like giant rafts of granite 'floating' on the denser basalt of the earth's inner crust (Fig. 82). There is a balance between the weight of the granite and how high it floats in the basalt. This state of balance is

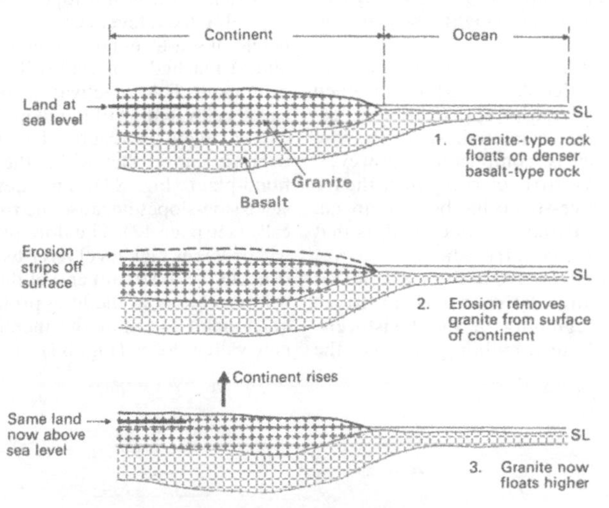

Figure 82 The principle of isostasy: a granite continent 'floats' on the denser basalt beneath. Removal of part of the continent by erosion causes the continent to float higher and sea-level around the continent appears to fall.

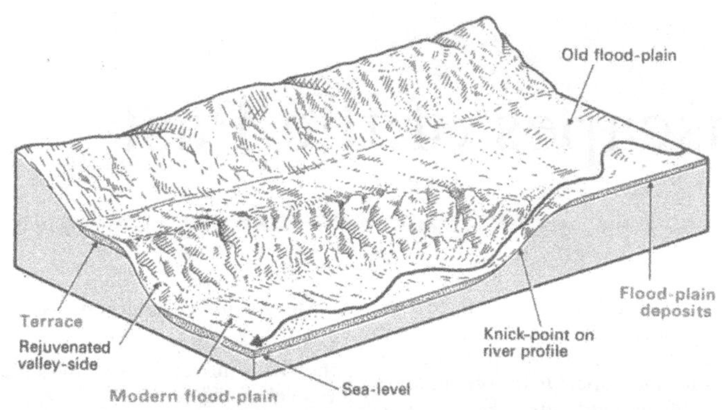

Figure 83 The rejuvenation of a river valley by a relative fall in sea-level causing an incision of the river, the formation of river terraces and a knick-point in the river profile.

termed **isostasy**. If part of the granite raft is removed by erosion, the granite will float higher on the basalt. This movement is called **isostatic re-adjustment**. A similar readjustment occurs if an enormous weight is added to the continent. During the Ice Age, for example, the weight of the ice-sheets caused Europe to sink slightly. Although that was a long time ago, parts of Europe (around the Baltic, for example) are still rising slowly as a result of the removal of that weight when the ice-sheets melted.

The level of water in the sea can also vary. In particular, during the Ice Age, sea-level dropped because water was being trapped in the ice-sheets. When the sheets melted, sea-level rose again. The situation is complicated, however, because land-level has also been rising since the Ice Age as the weight of ice-sheets has been removed. The result, as far as Britain is concerned, is that areas in the north of the country, where the ice was thickest, are still rising relative to the sea, whereas in the south, sea is rising relative to the land. Taken overall, since the beginning of the Pleistocene geological period, about 2 million years ago, the height of sea-level has varied from about 200 m above present sea-level to about 20 m below present sea-level.

1 *The results of a relative fall in sea-level*

In Chapter 2.D it was seen how rivers erode down towards sea-level. As they approach sea-level, vertical erosion becomes limited and the valley tends to widen because hillslope processes are eroding towards a fixed level (Chapter 2.E). If the *relative* height of sea-level falls, then the rivers which have almost reached sea-level will start to erode vertically again. This 'reactivation' of vertical erosion is called **rejuvenation** of the river. In the lower part of a river valley, the result will be that the river starts to erode a new valley within the floor of the former flood-plain (Fig. 83). This new valley will have steep side-slopes because the river is eroding vertically (see page 42). The floor of the old flood-plain will be left as high level **river terraces** on the valley side. At the upstream end of this new valley, there will be a break in the long profile of the river as a **knick-point** forms at the meeting of the old and new valley floors (Fig. 83).

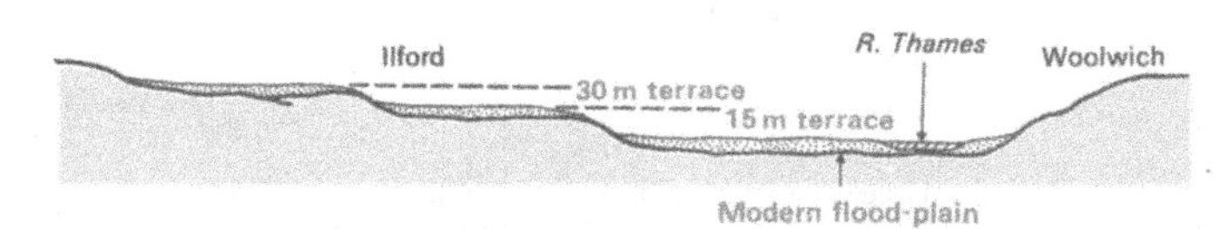

Figure 84 Terraces of the lower River Thames.

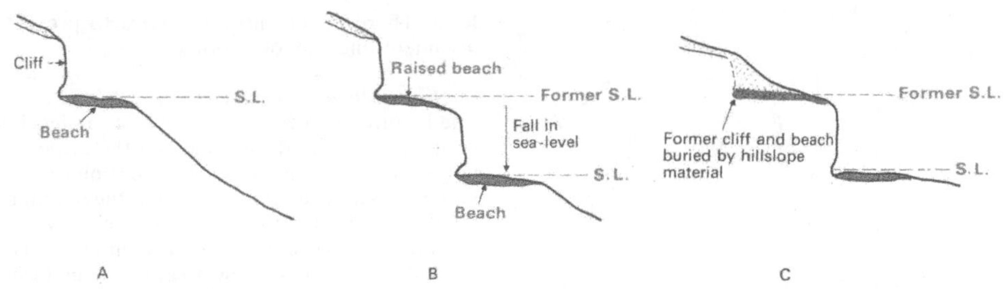

Figure 85 The formation of a raised beach by (B) a relative fall in sea-level and (C) its subsequent burial.

Because the relative height of land and sea is constantly changing, it is quite common to find rivers where rejuvenation has occurred on several occasions. Under these conditions, a whole series of terraces may be found on the valley side, with each terrace marking a former flood-plain level (Fig. 84).

When rejuvenation affects a meandering river, it may simply cause gradient to increase and the meanders to disappear, or the meander pattern may be preserved in bedrock if the knick-point retreats upstream along the line of the meanders. In the latter case, **incised meanders** are formed, producing a steep-sided winding valley like the lower part of the River Wye in Monmouth.

On the coast, a fall in sea-level will leave the original beach and cliff at some height above the new shoreline. Often, the beach has later become overgrown and the cliff has been eroded by normal hillslope processes so that only a slight notch appears in the modern slope profile (Fig. 85). In other places, however, the original feature is still clear and a **raised beach**, complete with sand and shells, can be seen. Around parts of the English coast, raised beaches can be found at about 8 m and 16 m above present sea-level. These heights sometimes correspond with the heights of river terraces above a modern river and beach and terrace may therefore have been formed at the same time.

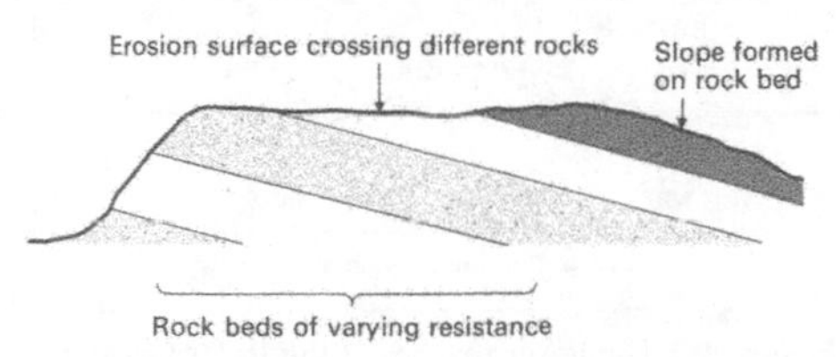

Figure 86 A high-level erosion surface crossing tilted rock beds of varying resistance.

Occasionally, much larger beach-like features can be found near the coast at some height above sea-level. These **benches** are often thought to be the remains of former wave-cut platforms.

Apart from benches, there are parts of the landscape, high above sea-level, which are almost level surfaces for many kilometres. Where these surfaces cross a number of different rock-types (Fig. 86), they must have been formed by erosion, but not by present-day processes. The general term **erosion surface** has been applied to these areas. In

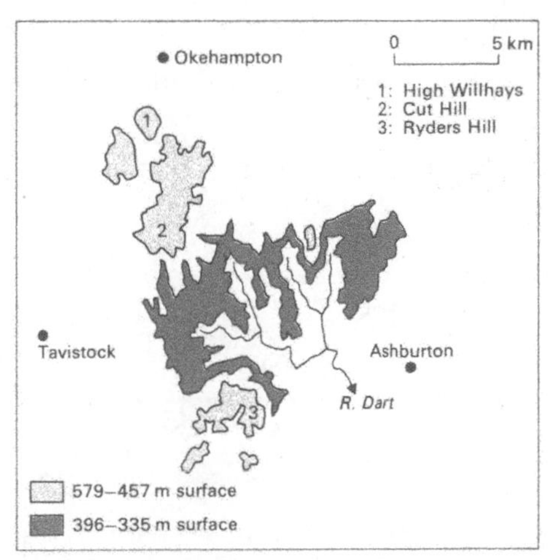

Figure 87 The remnants of two erosion surfaces in Dartmoor, Devon.

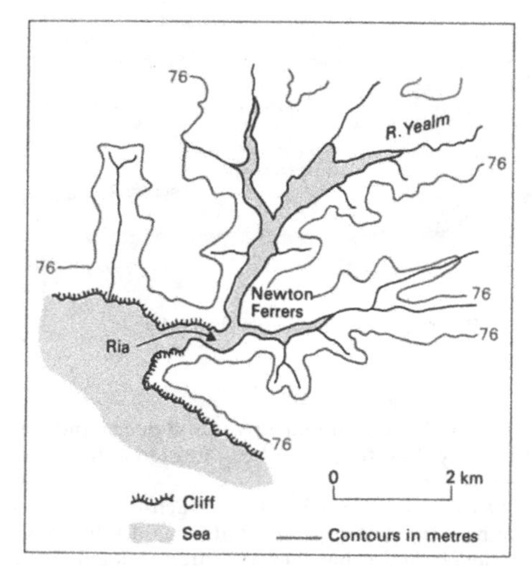

Figure 88 A ria, in South Devon, formed by the drowning of the valley of the River Yealm.

some parts of the country—South West England (Fig. 87), for example—the surfaces are fairly clearly defined. Elsewhere, the existence of a surface is very much a matter of speculation as a result of later erosion by river and hillslope processes. It is sometimes argued that these surfaces are really enormous wave-cut platforms, while it is also suggested that they may be the remnants of former river landscapes related to an earlier high sea-level. There is very little evidence to prove the argument one way or another.

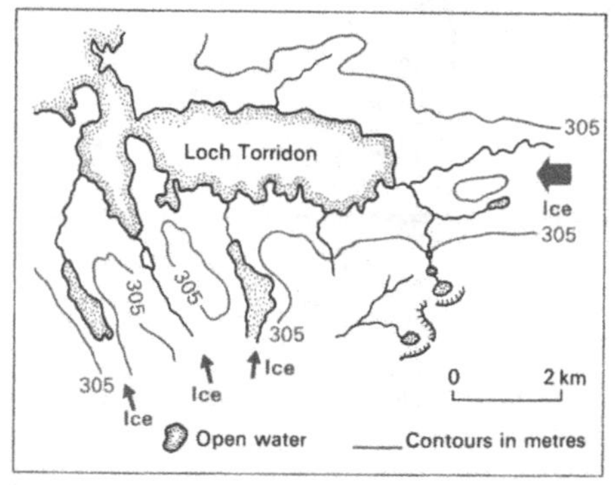

Figure 89 A fjord, Loch Torridon in Wester Ross, created by the drowning of a glacial trough.

2 *The results of a relative rise in sea-level*

The features resulting from a rise in sea-level are bound to be limited since most of the evidence is now buried beneath the sea. The one unmistakable feature of sea-level rise, however, is the flooding of the lower part of river valleys. Drowned river valleys are called **rias**. A map of a ria in Devon (Fig. 88) shows that the sea now occupies an inlet which winds and has tributaries just like a normal river valley. If a glaciated valley is later drowned by a relative rise in sea-level, the result is a **fjord**, common on the west coast of Scotland (Fig. 89) and Norway.

A drowned valley often provides an area of water sheltered from wave action, and, unless the tide is fairly fast, the inlet will soon fill with river and marine sediments. On the Sussex coast, for example, rivers such as the Cuckmere (Fig. 90) now meander over a plain which has been built of river and marine sediments after the lower valley was drowned and is therefore not a normal flood-plain.

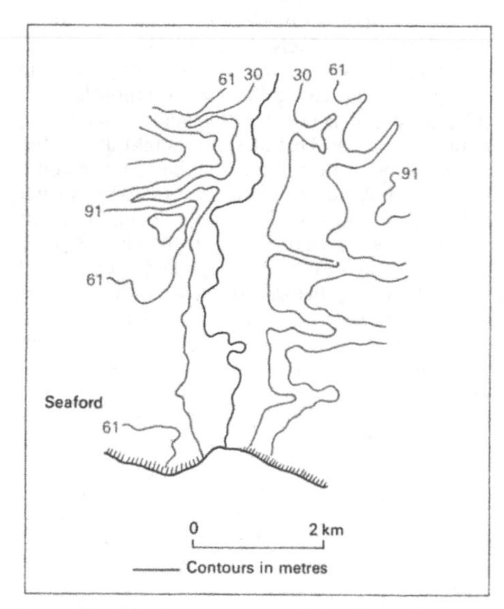

Figure 90 The lower reaches of the River Cuckmere, Sussex, where the 'flood-plain' is actually built of sediments deposited in a drowned river valley.

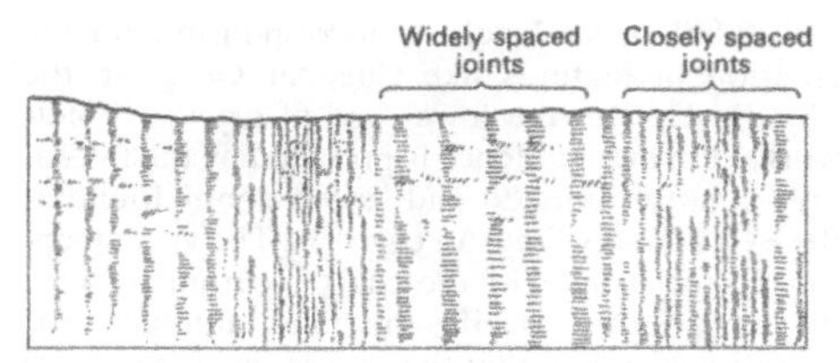

A Unweathered granite

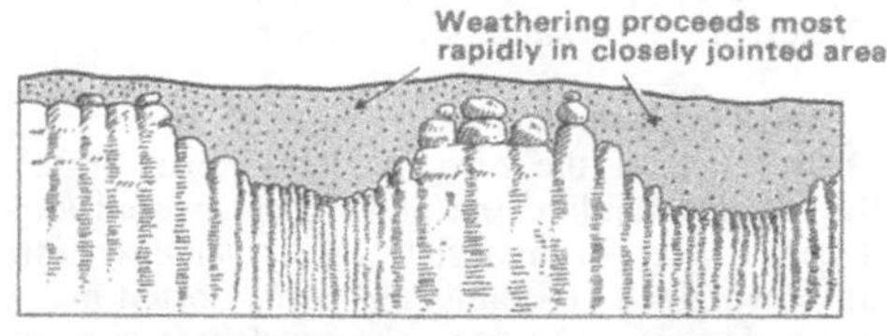

B Following deep tropical weathering

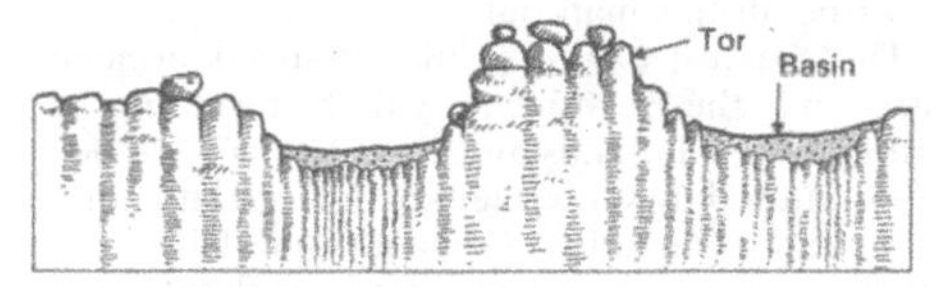

C Following removal of weathered material

Figure 91 The development of granite tors by the deep weathering of jointed granite under tropical conditions and the subsequent removal of weathered material.

B. Changes in climate

The earth's climate is in a constant state of change. The reasons for such changes appear to be enormously complex and linked largely to changes in the sun itself, although earth surface processes such as volcanic activity appear capable of altering the physical properties of the atmosphere. In addition, since the continental masses of the earth are in constant movement (**continental drift**), any point on the earth's surface will move through different climatic belts over geological time.

There is relatively little evidence in the surface landscape of Britain for earlier periods of tropical climate, although plenty of such evidence exists in the rocks of Britain. The best evidence is probably provided by locations like Dartmoor, in Devon, where the upstanding masses of bare granite, **tors**, lie above the general surface of the moor. The granite of Dartmoor is heavily jointed (Fig. 91A), but the joint pattern includes closely spaced and widely spaced areas. In the Mid-Tertiary geological period (about 25 million years ago), 'Dartmoor' was under the influence of a much warmer climate than that of today, which produced very deep weathering of the granite rock. (Very deep weathering, 30 m or more, can be found in the tropics today.) The depth of weathering, however, was not uniform, since the closer jointing encouraged more rapid weathering (Fig. 91B). The 'tors' were thus largely formed beneath the ground surface in the areas of widely spaced joints and were exposed by later erosion of the adjacent weathered material, probably by periglacial activity at a later period (Fig. 91C).

Far more obvious in the landscape today is the evidence of a past period of cold climate—the Ice Age. The Ice Age, in fact, covers the geological period called the Pleistocene, which started something over a million years ago. During the Pleistocene, there were actually at least four quite distinct **glacial periods** during which ice covered much of Northern Europe and North America. Each glacial lasted for up to 100 000 years and each is separated by an **interglacial** during which the climate at times became warmer than at present. There is some evidence to suggest that mankind is currently living in just another interglacial phase. It should perhaps be emphasised that within the pattern of four major glacials, there were countless minor readvances and subsequent retreats of ice, which makes any detailed analysis of the resulting landforms very difficult.

During each glacial period, ice seems to have accumulated first in the highland areas of Wales, Northern England and Scotland and then to have spread into the lowland, where it was joined by the sheet ice from continental Europe moving across the North Sea or the Irish Sea. As the ice advanced, so did the area of periglacial conditions in front of it. At maximum, the ice seems to have reached south to a line joining the Severn and Thames estuaries. South of this line, then, the Pleistocene was represented only by periglacial activity, whereas to the north the country was covered by successive ice-sheets and additional periglacial activity during ice-sheet retreat (Fig. 92).

1 *Periglacial features*

The effect of former periglacial activity is most marked on hillslopes. Many British scree slopes, such as those on the south side of Wastwater in Cumberland, were undoubtedly most active during the late Pleistocene as freeze–thaw cycles

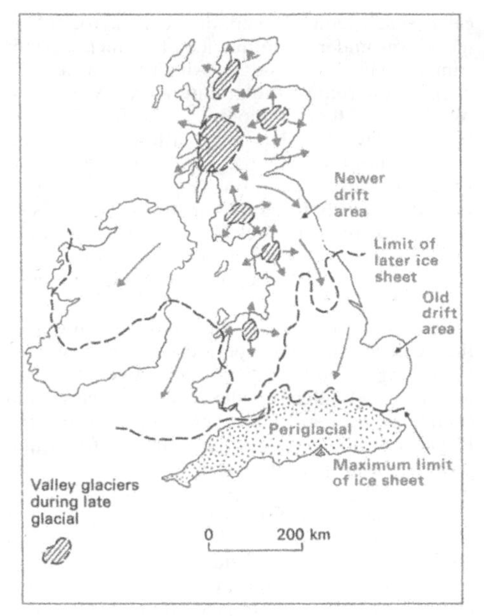

Figure 92 Former areas of glacial activity in the British Isles.

operated rapidly. Some screes in Southern England have since become totally inactive and vegetation-covered. Periglacial conditions favoured other mass movement processes such as landsliding. The old rotational landslides found inland from Lyme Regis, Dorset (Fig. 26) or on the escarpment of the Cotswold Hills were probably active in the Pleistocene as seasonal thaw produced saturated ground and very active rivers at the slope base. On a wider scale, solifluction seems to have been active on many slopes, producing a layer of poorly sorted angular debris beneath modern soils. This debris, called **head**, is usually absent from the upper part of hillslopes, but it accumulates in thickness towards the slope base and can often be seen exposed in river banks.

Normal river processes seem to have been more seasonal in nature, with channels and valleys therefore tending to develop in relation to a higher discharge than is found today. This is particularly noticeable in chalk and limestone areas, where dry valley networks appear to have been reactivated during the Pleistocene. Seasonal flow, coupled with a falling sea-level, seems responsible for the creation of features like Cheddar Gorge in the Mendip Hills at the lower end of dry valley networks. Further evidence is provided by the fans of only slightly rounded and sorted debris found at the exit of many limestone gorges. The very steep-sided short combs of the chalk escarpment were probably formed during this time, although it is sometimes argued that the process concerned was in the nature of concentrated solifluction (a 'mud-glacier') rather than a normal river.

2 *Ice-sheet features*

Over much of the lowland areas north of the Severn–Thames line, the landscape is covered with a thick layer of material which includes both unsorted glacial moraine and partly sorted melt-water deposits. The general term **drift** is used to describe all this material.

The greatest extent of the ice-sheets occurred early on in the Pleistocene, with the result that the drifts of the Midlands and East Anglia have been much altered by subsequent periglacial and humid temperate denudation. The landscape of the 'Older Drift' areas is therefore often without many

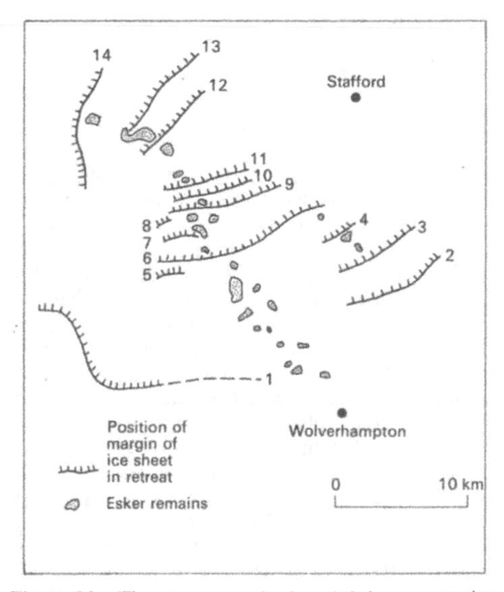

Figure 93 The remnants of eskers left by a retreating ice sheet in the Midlands plain to the north-west of Wolverhampton.

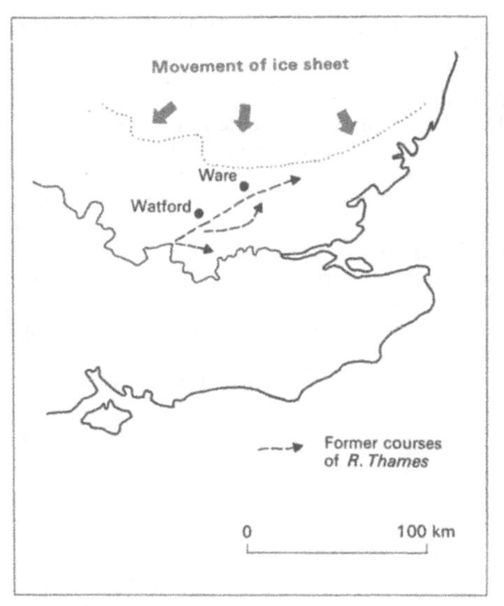

Figure 94 Successive courses of the River Thames during its diversion to the south by ice-sheets spreading from the north.

distinctive surface features. The preglacial valleys have been filled with drift and the former hills, such as the Chalk downs of Hertfordshire and Suffolk, have been lowered and rounded off by ice-sheet erosion.

In the lowlands farther north, the 'Older Drift' was completely obliterated by subsequent ice-sheets which left behind a 'Newer Drift' the features of which are far fresher. Within the 'Newer Drift' area, all the features described in Chapter 4 can be found. In the corridor between Wales and the Pennines, ice-sheets plastered the Cheshire Plain with ground moraine and drumlins. At the southern end of the corridor in Staffordshire, the edge of the retreating ice is marked by a series of great terminal moraines connected by the eskers formed in subglacial channels (Fig. 93).

Ice-sheets also had a very marked effect upon preglacial drainage patterns. In some cases, a river was simply diverted as the ice advanced. The River Thames, for example, originally flowed north-east through the sites of Watford and St Albans (Fig. 94). Progressive southward advances of the ice-sheet caused a series of shifts in course until the river adopted its current route.

In other cases, diversion was possibly produced as the result of ice blocking a river to form a

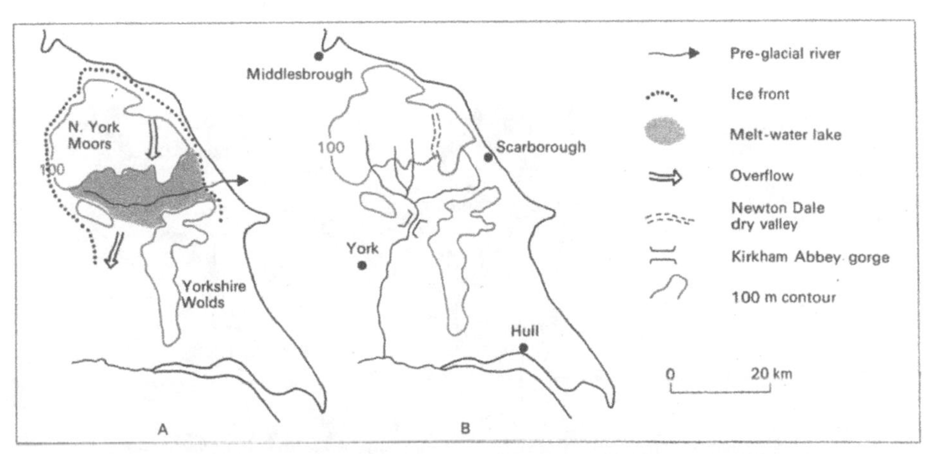

Figure 95 Alteration of drainage by ice in the North York Moors. Ice dammed water into lakes on north and south sides of the moors. The northern lake overflowed southwards through Newton Dale, which is now dry. In the south, Lake Pickering overflowed through Kirkham Abbey Gorge, which has become the new route for the River Derwent.

proglacial lake. In the North York Moors (Fig. 95A), a lake trapped against the north side of the moor by North Sea ice overflowed south, cutting the overflow channel of Newton Dale. In the post-glacial period, drainage on the north side of the moor reverted to its original pattern and Newton Dale was left as a high-level dry valley. Farther south, however, North Sea ice later blocked the eastern exit of the River Derwent, which became impounded on the site of what is now the Vale of Pickering and overflowed south-west, cutting a gorge at Kirkham Abbey (Fig. 95A). In this second case, post-glacial drainage has been fixed in this pattern: the western end of the Vale of Pickering is moraine-blocked and the Derwent continues to flow south-west into the Humber system (Fig. 95B).

3 *Valley glaciation features*

During the maximum extent of glaciation, the highland areas of Scotland, Wales and the Lake District were covered by ice-sheets which must have caused general erosion of the mountain tops. Towards the end of the glaciation, when the lowlands were free of ice, these mountains maintained local valley glaciers. It is these valley glaciers which have created much of the spectacular scenery of our mountains, and not the ice-sheets which preceded them.

In each area, it is the glacial trough which is the

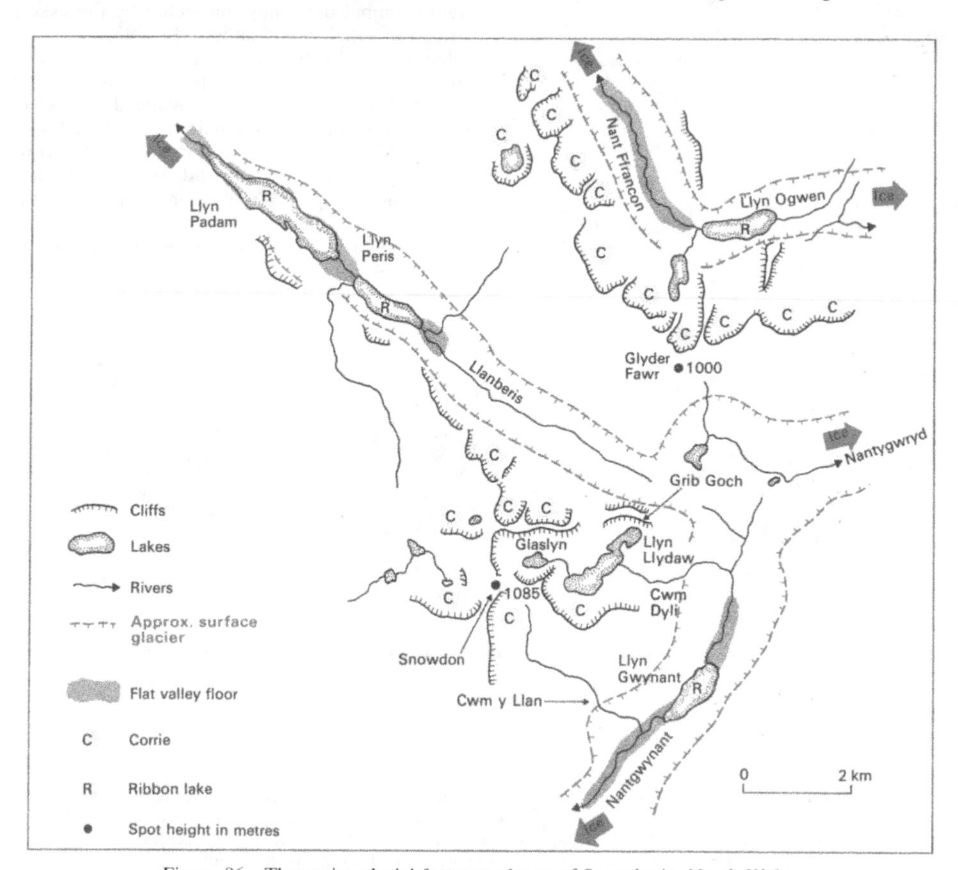

Figure 96 The major glacial features of part of Snowdonia, North Wales.

most impressive feature. In North Wales (Fig. 96), ice spread from Snowdon and the Glyder Mountains in a radial pattern. In most cases, the ice probably followed an already-existing river valley but proceeded to erode the enormous troughs of Nant Ffrancon, Llanberis and Nantgwynant. Each of these troughs is deep, straight and shaped rather like a 'U' in cross-section. Variations in trough width seem to occur frequently, and when width decreases, the depth may increase, leading to an overdeepened section which now holds a lake—for example, Llyn Gwynant. Other shallow lakes in the bottom of the trough, such as Llyn Padarn, are ribbon lakes which have been dammed by moraine. The glacial trough, of course, occupies only part of the modern valley in each case. The original upper surface of the glacier in each trough is marked as a bench on the sides of the modern valley. In the Llanberis valley, for example, the bench occurs at approximately 600 m altitude. Above that point, the steep smooth sides of the trough give way to lower-angle but rugged slopes.

Each glacier in the Snowdon system was fed by cirques high on the mountain sides which now form distinctive hollows, often occupied by a lake. Glaslyn, on the east side of Snowdon, is a cirque lake at the head of a hanging valley (Cwm Dyli) down which a tributary glacier must have flowed. Sharp arêtes, such as Grib Goch, are left between adjacent cirques, while the peak of Snowdon itself is a fretted dome created by steep cirque back-walls on three sides. It is noticeable in this area that the majority of cirques are facing towards the north and east (compare the two sides of the Glyder Mountains). South- and west-facing slopes were generally too warm for the formation of ice hollows. This tendency in itself indicates that much of the erosion took place at a time when the area was only *just* glacial.

Valley glaciers normally follow preglacial drainage routes and therefore serve largely to exaggerate the pre-existing landscape. Exceptions to this rule can be found, however, and one example occurs in the area to the west of Snowdon. Ice moving north and west from Snowdon (Fig. 97) would be flowing counter to the direction of the preglacial Nant-y-Betws valley. Some ice must have been diverted southwards down the Nant-y-Betws, but the rest accumulated in the north of the valley until it overflowed west into the Afon Llyfni valley and north into the Afon Gwyrfai. In each case, the watershed was breached and eroded by the ice until in the post-glacial period both the Afon Llyfni and Afon Gwyrfai start close to one

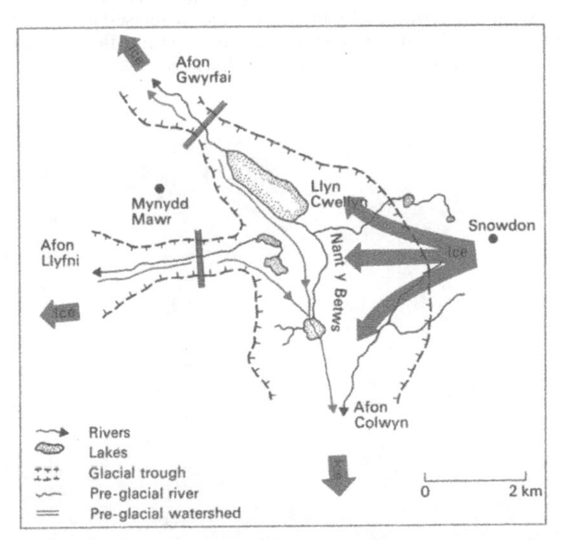

Figure 97 Interference in preglacial drainage patterns by ice moving westwards from Snowdon.

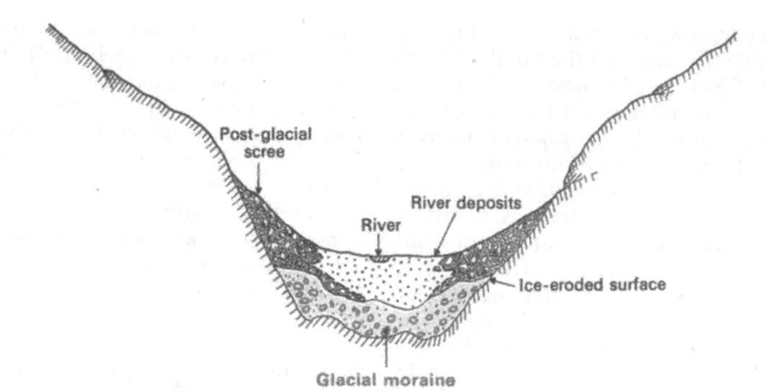

Figure 98 Cross-section through a former glacial valley showing some of the deposits which may cause alteration of the original trough shape.

another inside what used to be the Afon Colwyn's valley.

During ice retreat, glacial deposits were left in the upland areas. The trough base was often filled with ground moraine and successive positions of the glacier snouts are marked by terminal moraines, forming at times solid walls across the modern valley impounding ribbon lakes. Melt-water deposits are generally more difficult to identify, although various forms of kame and esker are quite common. In some cases, the depositional features have been substantially modified by later erosion, but very often they have been buried (Fig. 98) beneath periglacial scree and modern river deposits. In numerous cases, the apparently 'U'-shaped glacial trough turns out to be at least partly shaped by later processes.

Since glaciation, there has been a continuous modification of highland areas back towards 'typical' river landscapes. On the whole, in glaciated areas, rivers have a very steep headwater section before reaching the relatively low gradients of the main trough. Since river-long profiles are normally a fairly smooth curve, the post-glacial landscape often shows erosion of the upper mountain slopes by deeply incised gullies and deposition within the trough by fans, flood-plain deposits, lake deposits and scree slopes.

Further Reading

A. Books covering a wide range of examples

Sawyer, K. E., *Landscape studies* (Edward Arnold, 1975).
Small, R. J., *The study of landforms* (Cambridge University Press, 1970).
Strahler, A. N., *Introduction to physical geography* (John Wiley, 1973).

B. Books of further academic interest

1 *Theory*
Chorley, R. J. (ed.), *Water, earth and man* (Methuen, 1969).
Ward, R. C., *Principles of hydrology* (McGraw-Hill, 1974).
Weyman, D. R., *Runoff processes and streamflow modelling* (Oxford, 1975).
Morisawa, M., *Streams, their dynamics and morphology* (McGraw-Hill, 1968).
Leopold, L. B., Wolman, M. G. and Miller, J. P., *Fluvial processes in geomorphology* (W. H. Freeman, 1964).
Ollier, C., *Weathering* (Longman, 1969).
Young, A., *Slopes* (Longman, 1972).
Gregory, K. and Walling, D. E., *Drainage basin: form and process* (Edward Arnold, 1973).
Cooke, R. U. and Warren, A., *Geomorphology in deserts* (Batsford, 1973).
Embleton, C. and King, C. A. M., *Glacial and periglacial geomorphology* (Edward Arnold, 1968).
Price, R. J., *Glacial and fluvioglacial landforms* (Longman, 1973).
King, C. A. M., *Beaches and coasts* (Edward Arnold, 1973).

2 *Measurement techniques*
Hanwell, J. D. and Newson, M. D., *Techniques in physical geography* (Macmillan, 1973).
King, C. A. M., *Techniques in geomorphology* (Edward Arnold, 1966).
Gregory, K. J. and Walling, D. E., *Drainage basin: form and process* (Edward Arnold, 1973).
Weyman, D. R. and Wilson, C., *Hydrology for schools* (Teaching Geography Occasional Paper 25, Geographical Association, 1975).

3 *Applied geomorphology*
Cooke, R. U. and Doornkamp, J. C., *Geomorphology in environmental management* (Oxford, 1974).
Tank, R. W. (ed.), *Focus on environmental geology* (Oxford, 1973).
Chorley, R. J. (ed.), *Water, earth and man* (Methuen, 1969).
Steers, J. A. (ed.), *Applied coastal geomorphology* (Macmillan, 1971).

Index

Ablation 59–60, 65
Abrasion 35, 47, 55–6, 60, 64, 73
Accumulation zone 58–9
Alluvial fan 53–4, 67, 86
Alluvial plain 53
Angle of rest 27–8
Arch (marine) 75–6
Arete 61, 64
Arid landscapes 10, 24, 50–7
Avalanching 60

Bajada 53–4
Bar 78–9
Barchan dune 56–7
Baseflow 18
Basin of internal drainage 50
Bay 74–6
Bay-head beach 75, 77
Beach 77–9
Bedload 18
Bench 83
Bergschrund crevasse 64
Berm 77
Bornhardt 54
Boulder clay 64
Braided river 66–7

Canyon 53–4
Carbonate minerals 22
Catchment 15–16, 62
Cave 45–7; (sea) 75, 77
Chalk 23, 25–6
Chalk valley 48–9
Changes in sea level/land level 80–5
Channel (river) 33–40
Channel precipitation 14, 18
Chemical weathering 21–4, 51–2, 72–3
Cirque 61–4, 88–9
Clay minerals 22–3, 33–4
Clay soil 13–14
Cliff 73–4
Climate and denudation process 10–11, 39, 50–2, 58–60, 81, 85–6
Coastal landscapes 10, 24, 71–80, 83–4
Cohesion 27–8
Combe 49, 86
Consequent river 43–4
Constructional coastline 73, 76–80
Continental drift 85
Corrasion 35
Corrie 63–4, 88–9
Corrosion 24, 35, 72
Crag and tail 65
Crevasse 60–1, 64
Cross-profile (valley) 36, 42
Cwm 63–4
Cycle of denudation or erosion 12, 20, 42

Debris flow 52, 60, 69
Deflation hollow 57
Delta 66, 68, 80
Delta kame 66, 68
Dendritic drainage 43
Denudation 10; cycle 12, 20, 42; environment 11; in a river basin 38–44
Deposition 10; by ice 64–8; by river 37–8, 53–4, 80; by sea 76–80; in desert lake 54
Depression (limestone) 47–8
Desert 10–11, 24, 50–7; varnish 52
Differential erosion 74
Diffluent glacier, 63, 89
Dip slope 44
Discharge (riverflow) 15–20; and downstream changes 35–7; and transportation 34; of desert rivers 50; of periglacial rivers 69
Dissolved load of river 33, 39–41
Dissolved minerals 22–3, 25
Distributary channel 80
Diversion of river 87–8
Doline 47
Downstream changes 35–8
Draa 57
Drainage basin 15–16, 19, 38–44
Drainage pattern 43–4
Drift 86
Drowned valley 84
Drumlin 64–5, 68, 87
Dry valley 48–9
Dune (desert) 56–7
Dynamic equilibrium 42

Earth movement 9, 11–12
Ephemeral river 50
Equal-energy distribution 35
Equilibrium in landscape 32, 42
Erg 55
Erosion 10; by ice 60–4, 68; by river 35–8; by wind 55, 57
Erosion surface 83–4
Erosional coastline 73–6
Erratics 68
Escarpment 44
Escarpment retreat 49
Esker 66, 68, 86–7
Estuary 71, 80
Evaporation 13, 17, 50
Evaporite deposit 54
Evapotranspiration 16, 18

Fault in rock 75
Feldspar 12, 22–3
Fetch 71–2, 75
Firnline 59
Fjord 84
Flash flood 50
Flood flow 18–19, 38
Floodplain 38–9, 82
Flow duration curve 19, 39
Freeze-thaw weathering 21–2, 51, 60, 64, 69, 85
Friction (between particles) 27–8
Friction (in a channel) 37
Frozen soil 69–70

Geological cycle 10
Geological structure 10; and denudation 39, 43–4, 52, 74–6
Geology and water 19

Glacial landscapes 10, 24, 58–70, 86–90
Glacial deposition 64–5, 67, 86, 90
Glacial diversion of drainage 87–8
Glacial meltwater 59–60, 66–8, 87–8
Glacial trough 61–4
Glacier 58–68, 88–90
Gorge 48–9, 53
Granite 11–12, 23; tor 85
Ground moraine 61, 64–5, 68, 87
Ground water 15–16, 18–19, 25, 45, 47
Groyne 77
Gully 26

Hanging valley 61, 63, 89
Head 86
Headland 74–6
Headwards erosion 43–4
Hillslope 25–33, 52–4, 60; in relation to river 41–2
Humid landscapes 10, 13–49
Hydraulic action 35, 73
Hydrograph 18–21
Hydrology: of arid areas 50; of glaciers 58–60; of limestone areas 45; of river systems 13–21

Ice Age 49, 85
Ice sheet 12, 58, 68, 82, 85–8
Igneous rock 11–12, 22–3
Incised meander 83
Infiltration 13–14, 19
Inselberg 53–4
Insolation weathering 51
Interception 13
Interglacial 85
Interlocking (particles) 27
Intermittent river 50
Isostasy 81–2

Kame 66, 68
Kame terrace 66, 68
Kettle hole 67
Knick-point 82

Lagoon 78–9
Lake: desert 50, 54; glacial 63, 87–9; proglacial 66–7, 87–8
Lake plateau 68
Landslide 28–9, 40, 86
Land use and runoff 19
Lateral beach 78
Lateral erosion 36
Lateral moraine 61, 65, 67
Levee 38
Limestone 11–12, 22–6, 33, 44–9
Limestone pavement 45, 49
Loess 57
Long profile (river) 35, 90; glacier 62–3
Longitudinal dune 56–7
Longshore drift 77–9

Marine deposition 76–80
Marine erosion 71–6
Mass movement 27–31, 33, 40–1, 52, 60, 69, 86
Meander 36–9, 83
Meltwater 59–60, 66–8, 87–8
Mesa 53–4
Moraine 60–1, 64–8, 87
Mudflat 80
Mudflow 29–30, 40, 86

Nivation hollow 64

Obsequent river 43–4
Ocean current 71
Off-shore bar 78–9
Outwash plain 66–7
Overdeepening 63
Overflow channel 67, 87–8
Overland flow 14, 18–20, 26–7, 32–3, 40, 52
Ox-bow lake 38–9

Particle load of river 33–4, 39–41
Patterned ground 69
Pediment 53
Percolation 15–16, 45
Percolation zone (limestone) 47
Periglacial landscapes 10, 58, 68–70, 85
Permafrost 69–70
Permeability 13–14, 16, 49
Pheatic cave 47
Pingo 70
Pipe, soil 14–15
Plateau 48, 83
Playa lake 54
Plucking (ice) 60, 63–4
Plunge pool 36
Porosity 13
Post-glacial landscape 89–90
Pothole (river) 36
Pothole (cave) 45–7
Precipitation13, 17–18, 50
Proglacial lake 66–7, 87–8
Push moraine 65
Pyramidal peak 64

Quartz 12, 23

Raised beach 83
Recurved ridge 78
Refraction, wave 74–5
Rejuvenation 49, 82–3
Resurgence 45–6
Rhourd 57
Ria 84
Ribbon lake 88–9
Riffle (river) 35
Rill 26
Ripple 57
River capture 43–4
River deposition 33, 37–8 90
River erosion 33, 35–7
River long profile 35
River meander 36–7
River terrace 82
River transportation 33–5
River valley 36, 42
Riverflow 13, 15–20
Riverflow velocity 33
Roche moutonee 63
Rock basin 63
Rockfall 27, 49
Rock permeability 16
Rock type and resistance 11
Rotational landslide 28–30, 86
Runoff 13, 16, 18, 20–1

Salt crystal growth 52
Salt lake 54
Salt marsh 80
Saltation 34, 55
Sand dune 56–7
Sand sea 55
Sandy soil 13–14
Saturated rock/soil 15–16
Scree 27–8, 49, 52, 60, 69, 85, 90
Sea-level change 79–80, 80–5
Sediment (river) 33–8
Sedimentary rock 11–12, 22–3
Semi-arid landscapes 10, 50–7
Sheetflow 52
Sief dune 56–7
Silicate minerals 27
Sink hole 47
Slope, *see* hillslope
Snout (glacier) 59, 65
Snow 58–9
Snowfield 61
Soil creep 30–2, 40
Soil erosion 26–7
Soil formation 24–5
Soil particles 13–14
Soil permeability 13
Soil pipe 14–15
Soil porosity 13–14
Soil type and runoff 19
Soil water 13
Solifluction 69, 86
Solutional weathering 22, 45–7
Sorting of particles 28, 38, 64, 79
Spit 77–8

Spring 15, 45
Stack 75
Steady state 42
Stalactite/stalagmite 47
Stone pavement 52
Stone net 69
Stone stripe 69
Storm hydrograph 18
Storm ridge 77
Strength (of material) 27
Stress (applied) 27
Striation 60
Subglacial channel 66–8
Subsequent river 43–4
Surface runoff, *see* overland flow
Suspended load of river 34–5, 52
Swallet/swallow hole 45–7

Terminal moraine 61, 65, 87
Terrace, river 82
Terracette 30–1
Throughflow 14, 18–19, 25, 33, 40
Tidal scour 71
Tide 71
Till 64, 68
Time (as a geological factor) 12, 39–41
Tor 85
Transpiration 13
Transportation 10; by glaciers 60; by hillslope processes 25–31, 52; by ice sheets 68; by rivers 33–5, 38; by waves 76–80; by wind 54–5, 70
Transport-limited slope 31
Transverse dune 56–7
Trellis drainage 43–4
Tropical climate and denudation 85
Truncated spur 62

U-shaped valley 61–2, 90

Vadose cave 47
Valley, dry 48–9
Valley, glacier 58–64, 88–90
Valley, river 25, 35–6, 41–2
Varve 67
Velocity: of glaciers 58–9; of overland flow 26; of riverflow 33–8; of soil creep 30, 32; of tides 71
Ventifact 55
Vertical erosion 35–6, 42, 82
Volcanic activity 9

Water balance 16–17, 58–9; equation 17
Water cycle 13–16
Waterfall 36
Watershed 16, 62–3
Water table 15–16, 45, 49
Wave 71–3
Wave-cut notch 73–4
Wave-cut platform 74
Wave erosion 72–3
Weathering 10; chemical 21–5, 33, 51–2, 60, 72–3; physical 21–2, 24–5, 51–2, 60, 62, 64, 68
Wind deposition 56–7
Wind erosion 55
Wind transportation 55

Yardang 55